T0135967

Diss. ETH No. 17259

A SYSTEM CONCEPT FOR ULTRA WIDEBAND (UWB) BODY AREA NETWORKS

A dissertation submitted to the

SWISS FEDERAL INSTITUTE OF TECHNOLOGY

ZURICH

for the degree of

Doctor of Sciences

presented by

THOMAS ZASOWSKI

Dipl.-Ing., Saarland University

born December 14, 1975

citizen of Germany

accepted on the recommendation of

Prof. Dr. A. Wittneben, examiner

Prof. Dr. D. Dahlhaus, co-examiner

Dr. S. Launer, co-examiner

2007

Reihe Series in Wireless Communications
herausgegeben von:
Prof. Dr. Armin Wittneben
Eidgenössische Technische Hochschule
Institut für Kommunikationstechnik
Sternwartstr. 7
CH-8092 Zürich

E-Mail: wittneben@nari.ee.ethz.ch
Url: http://www.nari.ee.ethz.ch/

Bibliografische Information der Deutschen Nationalbibliothek

Die Deutsche Nationalbibliothek verzeichnet diese Publikation in der
Deutschen Nationalbibliografie; detaillierte bibliografische Daten sind
im Internet über http://dnb.d-nb.de abrufbar.

ISBN 978-3-8325-1715-1
ISSN 1611-2970

Logos Verlag Berlin GmbH
Comeniushof, Gubener Str. 47,
10243 Berlin
Tel.: +49 030 42 85 10 90
Fax: +49 030 42 85 10 92
INTERNET: http://www.logos-verlag.de

Unser Kopf ist rund, damit die Gedanken die Richtung ändern können.

–Francis Picabia

Danke

Ich möchte mich an dieser Stelle bei all denen bedanken, die zum Gelingen dieser Arbeit beigetragen haben.

Mein besonderer Dank gilt Prof. Dr. Armin Wittneben. Ihm möchte ich für dieses interessante und herausfordernde Thema sowie für die wissenschaftliche Unterstützung danken.

Meinen Korreferenten Prof. Dr. Dirk Dahlhaus und Dr. Stefan Launer danke ich für die Begutachtung meiner Dissertation und für die hilfreichen Kommentare.

Für seinen Rat bei Fragen in den Bereichen Wellenausbreitung und Antennen möchte ich mich bei Dr. Gabriel Meyer bedanken. Dem UWB-Team, bestehend aus Dr. Frank Althaus, Florian Trösch und Christoph Steiner, möchte ich für die vielen anregenden Diskussionen danken. Ihnen und meinen Kollegen Etienne Auger, Stefan Berger, Celal Esli, Azadeh Ettefagh, Dr. Ingmar Hammerström, Dr. Marc Kuhn, Dr. Carlo Mutti, Oliver Nipp, Georgios Psaltopoulos, Dr. Boris Rankov, Jörg Wagner und Jian Zhao gilt mein Dank auch für die gute Zusammenarbeit und für das angenehme Klima am Institut, das sie insbesondere durch zahlreiche Ski- und Wanderausflüge sowie Fussballmatches gefördert haben.

Barbara Aellig, Dr. Jürgen Kemper, Tobias Meier und Claudia Zürcher möchte ich für die Unterstützung bei administrativen Aufgaben danken.

Hansruedi Benedickter möchte ich für die Unterstützung bei den Messaufbauten danken.

Für die gute Zusammenarbeit und viele interessante Diskussionen möchte ich mich bei Stefan Hänggi, Dr. Stefan Launer und Dr. Herbert Bächler von der Firma Phonak bedanken.

Ganz besonders möchte ich meinen Eltern danken, die mich während meiner gesamten Ausbildung gefördert haben.

Mira danke ich sehr, dass sie mir immer zur Seite gestanden und damit massgebend zum Gelingen dieser Dissertation beigetragen hat.

3

Abstract

Wireless body area networks have recently gained a lot of interest due to multiple possible applications such as wireless health monitoring or wearable computing. Because of the rather simple hardware realizations and the energy efficiency, ultra wideband (UWB) communication has become one promising technology for the use in wireless body area networks (BAN). After pointing out the motivation of this work and highlighting its contribution, a definition of body area networks is presented as well as a brief introduction on UWB. There, the main promises of UWB communications are presented as well as the principles of some typical receiver structures for UWB.

Since UWB communication at the human body is a brand new topic, channel measurements at the human body are performed. The frequency range for these measurements is chosen from 2 to 8 GHz. Based on 1100 channel measurements a channel model for the UWB BAN is derived. Using the Akaike information criterion (AIC) it is shown that the channel decays over the time and that the channel taps are log-normal distributed. The channel at the head is of particular interest as most human communication organs such as mouth, ears, and eyes are located there. Therefore, the ear-to-ear link, which can be regarded as a worst case scenario at the head due to the missing line-of-sight component, is considered to specify the impact of the channel on the system design. When considering the ear-to-ear link it is shown by means of theory, simulations, and measurements that the direct transmission through the head is attenuated so much that it is negligible. Therefore, antennas should be designed in a way that they do not radiate into the body but away from it or along its surface. Moreover, it is shown that the channel is robust against distance variations between the antenna and the skin, and that reflections and absorptions are caused by the body. For the ear-to-ear link the antennas should be placed behind the ears to get the smallest channel attenuations. From the measurements it can also be observed that the main energy of the channel impulse response is contained in a very small time window. Thus, non-coherent receivers with a short integration duration can capture almost the whole energy of the channel.

Since UWB systems are a secondary spectrum user, the impact of existing wireless services on UWB is investigated as well. Due to the low transmit power not only the in-band but also the out-of-band interferers are harmful for UWB transmission. Based on frequency-domain and time-domain measurements it is shown that interference not close to the UWB device can be handled by using filters. However, this is not sufficient enough if an interferer is in close vicinity of the UWB device. Therefore, the temporal cognitive medium access is presented to avoid the interference from burst-wise transmitting devices. There, the UWB system listens if the channel is occupied by an interferer and it transmits only in case that no other system is active at the same time. For such a temporal cognitive MAC an expression is given to calculate the optimum UWB packet length. Assuming different interference scenarios, the packet lengths are evaluated. Moreover, it is shown that reasonable usable idle times can be achieved, which the UWB device can use for transmission, and strict latency time requirements can be met. ALOHA, 1-persistent CSMA, and non-persistent CSMA are considered as access schemes for the performance evaluation of the temporal cognitive MAC. For evaluation, two different cases are distinguished, with and without bandpass filter at the UWB receiver. It is shown that a UWB device with bandpass filter that uses the temporal cognitive MAC in conjunction with non-persistent CSMA has low packet error rates below 10^{-2} for up to about 15 active UWB links.

Due to complexity reasons non-coherent receivers are the most promising solution for the use in UWB devices. Hence, the focus in this thesis lies on the energy detector and the transmitted reference receiver, which have both the same performance. Furthermore, the maximum likelihood receivers in the presence of inter-symbol interference are derived for binary pulse position modulation and transmitted reference pulse amplitude modulation assuming partial channel state information. The maximum likelihood receivers in the presence of a co-channel interference are calculated for the transmitted reference PAM as well. A family of maximum likelihood receivers is also derived for the transmitted reference pulse interval amplitude modulation, which is a combination of pulse position modulation and transmitted reference pulse amplitude modulation. The performance of all these receiver structures is evaluated by means of bit error rate simulations. The simulations are performed by using channels with independent and identically-distributed channel taps and exponential decaying channels as well as by using the BAN channel model. For all these receiver families a trade-off between performance and complexity is observed. Assuming a higher level of channel state information the performance improves while the complexity increases. The receiver structures with knowledge of the average power delay profile are recommended for the use in wireless BAN. These receiver structures exhibit for most channels better performance than the ones without channel state information, however, they require only moderately higher complexity. Furthermore, the receivers with knowledge of the average power delay profiler are less sensitive to the chosen integration duration, since the weighting can be regarded as choosing a variable integration duration.

Finally, recommendations for a UWB BAN system are given and conclusions are presented.

Abstract

Kurzfassung

Aufgrund einer Vielzahl möglicher Anwendungen, wie z.b. der drahtlosen Patientenüberwachung oder in Kleidungsstücken integrierten Computern, haben drahtlose Körperbereichsnetze in der letzten Zeit eine grosse Beachtung erfahren. Ultrabreitband (UWB, engl. Ultrawideband)-Kommunikation ist eine vielversprechende Technologie für die Verwendung in Körperbereichsnetzen, weil Schaltungen relativ einfach realisiert werden können und zudem noch energieeffizient sind. Beginnend mit einer Motivation dieser Arbeit wird anschliessend eine Definition zu Körperbereichsnetzwerken präsentiert und eine kurze Einleitung in UWB gegeben. Dabei werden sowohl die besonderen Eigenschaften von UWB als auch einige typische UWB-Empfängerstrukturen präsentiert.

UWB-Kommunikation am menschlichen Körper ist ein ganz neues Forschungsgebiet. Aus diesem Grund werden Kanalmessungen am menschlichen Körper durchgeführt. Der Frequenzbereich für diese Messungen ist von 2 bis 8 GHz gewählt. Basierend auf 1100 Kanalmessungen wird ein Kanalmodell für UWB Körperbereichsnetzwerke entwickelt. Unter Verwendung des Akaike Informationskriteriums (AIC, engl. Akaike Information Criterion) wird gezeigt, dass die Abtastwerte der Kanalimpulsantworten log-normal verteilt sind. Der Kanal am menschlichen Kopf ist von besonderem Interesse, da sich die meisten menschlichen Kommunikationsorgange, wie z.B. Mund, Ohren und Augen, am Kopf befinden. Deswegen wird die Ohr-zu-Ohr Verbindung, welche aufgrund der fehlenden direkten Komponente als ungünstigster Fall gesehen werden kann, verwendet, um den Einfluss auf den Systementwurf zu spezifizieren. Für die Ohr-zu-Ohr Verbindung wird mittels Theorie, Simulation und Messungen gezeigt, dass der

9

direkte Pfad durch den Kopf so stark gedämpft ist, dass er vernachlässigt werden kann. Aus diesem Grund sollten Antennen so konstruiert werden, dass sie nicht in den Kopf strahlen, sondern dass die Strahlung entweder von ihm weg oder entlang seiner Oberfläche erfolgt. Ausserdem wird gezeigt, dass der Kanal robust gegenüber unterschiedlichen Distanzen zwischen Antenne und Haut ist und dass Reflexionen und Absorptionen durch den Körper verursacht werden. Für die Ohr-zu-Ohr Verbindung sollten die Antennen hinter den Ohren platziert werden, um die Dämpfung möglichst gering zu halten. Anhand der durchgeführten Messungen kann ausserdem beobachtet werden, dass der Hauptteil der Energie einer Kanalimpulsantwort in einem sehr kurzen Zeitfenster enthalten ist. Somit können inkohärente Empfängerstrukturen mit einer kurzen Integrationszeit fast die gesamte Energie sammeln, welche im Kanal vorhanden ist.

Da UWB Systeme einen Zweitnutzer des Spektrums darstellen, wird auch der Einfluss bereits existierender drahtloser Systeme auf UWB untersucht. Wegen der niedrigen UWB-Sendeleistung sind nicht nur Störer im gleichen Frequenzband kritisch, sondern auch Störer ausserhalb dieses Bandes. Basierend auf Zeit- und Frequenzbereichsmessungen wird gezeigt, dass Störer, die sich nicht in direkter Nähe des UWB Empfängers befinden, durch ein Filter unterdrückt werden können. Dies genügt jedoch nicht bei Störern in direkter Nähe des Empfängers. Um Störungen von paketweiseübertragenden Systemen zu reduzieren, wird die sog. „temporal cognitive" Kanalzugriffskontrolle (MAC, engl. Medium Access Control) eingeführt. Dabei hört das UWB System, ob der Kanal von einem Störer genutzt wird und sendet nur im Fall, dass kein anderes System gleichzeitig aktiv ist. Für ein solches Kanalzugriffsverfahren werden die optimalen Paketlängen sowie die optimalen UWB Paketlängen für unterschiedliche Interferenzszenarien bestimmt. Es wird gezeigt, dass bei Verwendung dieses Kanalzugriffsverfahrens strikte Latenzzeiten eingehalten werden können. ALOHA, 1-persistent CSMA und non-persistent CSMA werden bei der Evaluierung des „temporal cognitive" MAC als Zugriffsverfahren verwendet. Weiter werden die beiden Fälle berücksichtigt, dass Störungen im UWB Empfänger nicht gefiltert werden und dass ein Bandpassfilter verwendet wird. Es wird gezeigt, dass ein UWB Gerät, welches non-persistent CSMA verwendet, eine Paketfehlerrate von weniger als 10^{-2} aufweist, falls nicht mehr als fünfzehn andere UWB Verbindungen aktiv sind.

Aufgrund der geringen Komplexität sind inkohärente Empfänger besonders attraktiv für UWB

Systeme. Deswegen liegt der Fokus dieser Arbeit auf dem Energiedetektor und dem sog. Transmitted-Reference Empfänger, welche beide die gleiche Leistungsfähigkeit zeigen. Des Weiteren werden Maximum-Likelihood Empfänger unter Berücksichtigung von Intersymbolinterferenz für binäre Pulspositionsmodulation sowie für Transmitted-Reference Pulsamplitudenmodulation hergeleitet. Dabei wird auch ein unterschiedlicher Grad an Kanalzustandsinformation berücksichtigt. Für Transmitted-Reference Pulsamplitudenmodulation werden auch die Maximum-Likelihood Empfänger in Gegenwart eines gleichartigen Störers berechnet. Zusätzlich wird für Transmitted-Reference Pulsintervalamplitudenmodulation eine Familie von Maximum-Likelihood Empfängern hergeleitet. Die Leistungsfähigkeit dieser Empfängerstrukturen wird mittels Bitfehlerraten ausgewertet. Die Simulationen werden für drei verschiedene Arten von Kanälen durchgeführt. Dies sind Kanäle mit unabhängigen, gleichverteilten Abtastwerten der Kanalimpulsantwort, exponentiell abfallende Kanäle sowie das für den menschlichen Körper erstellte Kanalmodell. Für alle Empfängerfamilien kann ein Kompromiss zwischen Leistungsfähigkeit und Komplexität beobachtet werden. Mit höherer Kanalzustandsinformation nehmen sowohl die Leistungsfähigkeit als auch die Komplexität zu. Die Empfängerstrukturen mit Kenntnis des mittleren Leistundsdichtespektrums werden für den Einsatz in Körperbereichsnetzen empfohlen. Diese Empfängerstrukturen zeigen für die meisten Kanäle eine bessere Leistungsfähigkeit als die Empfängerstrukturen ohne Kanalzustandsinformation, benötigen aber nur geringfügig höhere Komplexität. Ausserdem sind diese Empfänger weniger sensibel gegenüber der gewählten Integrationsdauer, da die durchgeführte Gewichtung als variable Integrationsdauer betrachtet werden kann.

Schlussendlich werden Empfehlungen für ein UWB System für Körperbereichsnetzwerke ausgesprochen und eine Zusammenfassung präsentiert.

Contents

IV Physical Layer 121

V Medium Access Control 191

VI Conclusions 213

Contents

Part I

Introduction

Chapter 1

Motivation

In the recent years miniaturization, energy efficiency, and cost reduction of electronic devices have made a rapid progress. Hence, a number of new applications have appeared and portable electronic devices have been playing an important role in our everyday life. Nowadays, almost everyone owns a mobile phone and many persons use notebooks and personal digital assistants (PDAs). The sale numbers of multimedia players and digital cameras have also increased substantially over the last few years. Thus, almost every person carries several electronic devices with her or with him. In the future, this number will even increase having prospective applications in mind such as health monitoring, video glasses, or sensors and actuators in the clothes that assist the person wearing them. Depending on the scenario, there might exist nodes with different capabilities resulting in a very heterogeneous structure. In this type of scenario, some devices have only limited capabilities and generate little data to transmit, while other devices require high data rates having no strict energy limits. However, there exist also scenarios, where all nodes have the same capabilities yielding a homogenous network topology. Anyway, portable devices are only rarely mutually connected by wireless technologies or cables. The existing solutions have some major drawbacks. Either they are inconvenient to use (cables), support only low data rates (existing wireless technologies such as Bluetooth), or suffer from both (infrared). Hence, portable devices can benefit from a new energy efficient wireless solution offering higher data rates than the existing wireless solutions.

One promising wireless technology for such a group of electronic devices, which is referred to as wireless body area network (WBAN), is ultra-wideband (UWB) communication. UWB offers a number of distinct features compared to narrowband systems. Due to the very wide bandwidth from several hundred MHz up to a few GHz, UWB allows for very high data rates [1]. Nevertheless, UWB is very well suited for heterogeneous networks since data rates are easily scalable [2]. The very wide bandwidth results in transmission of very short pulses in time domain. Since there are theoretically no mixers or phased-lock loops (PLLs) necessary for pulse transmission, UWB is claimed to allow for simple hardware realizations [3]. Moreover, a very low power consumption of UWB is expected [4] due to the low complexity of the hardware and the very low allowed transmit power. This low transmit power and the short pulse structure of the UWB signals lead to the assumption that UWB systems are robust against the interference from other UWB systems as well as from narrowband systems [5]. Although the transmit range of UWB is limited by the low transmit power, this limitation reduces the electromagnetic exposure of the body and hence is likely to increase the user acceptance of wireless transmission close to the body.

Chapter 2

Contribution and Outline

UWB communication offers a number of distinct features such as energy efficiency or simple hardware realizations. Therefore, UWB is considered as a promising candidate technology for the communication in wireless body area networks (WBANs). In this thesis, we investigate in particular the UWB WBAN channel and the non-coherent receiver structures with low complexity. Additionally, a MAC principle is presented which avoids interference from burst-wise transmitting non-UWB systems.

Until a few years ago, no investigations on UWB in WBANs have been made. In 2002, measurements in the frequency range 1-11 GHz were performed with only one antenna mounted on the body [6] showing a deep notch in the receive signal pattern caused by the body. Shortly after this work we presented the first channel measurements with both antennas mounted on the body [7], followed by [8] where channel measurements were performed in an indoor environment. It was shown that the frequency correlation properties of the channel depend on the considered frequency band and that the signal energy spread in time-delay domain varies significantly. In [9], the BAN channel in the frequency range 2-6 GHz was investigated by means of Finite Difference Time Domain (FDTD) simulations. A path loss model was derived and a power delay profile as well as an amplitude distribution were evaluated. The simulation results were verified by measurements in a parking lot. The same authors observed an impact of the arm motion on the receiver power and ground reflections that could assist communication from one side of the body to the opposite [10]. The authors extended their work and presented a channel model in the frequency range 3-6 GHz based on 144 measurements [11]. Path loss at the human body were evaluated in [12] for the frequency range 3-9 GHz using two different antennas.

To investigate the impact of the UWB BAN channel on the communication system, we performed measurements in the frequency range from 2 to 8 GHz. Based on these measurements, a simple time-domain UWB BAN channel model is derived considering 20 different links at the body yielding 1100 measurements altogether. By using the Akaike information criterion (AIC) it is shown that the channel taps are log-normal distributed. Since the AIC only determines the most likely distribution out of a set of beforehand given distributions, the log-normal distribution is verified by comparing the cumulative distribution functions. Besides the distribution of the channel taps also the power delay profile and the path loss are determined. Due to the proximity of the most important human communication organs such as mouth, eyes, and ears, the head is very attractive for the placement of transmitters and receivers. Focusing on the ear-to-ear link, which can be regarded as a kind of worst case scenario for the communication at the head, the impact of the channel characteristics on the communication system is discussed. By means of measurements, simulation, and theory it is shown that the attenuation by the head is so large that there exists no direct link through the head. Hence, an antenna for a BAN should be designed in a way that it does not radiate into the body. Moreover, it is shown that the channel is robust against a varying distance between antenna and skin. Although the channel changes significantly for different antenna positions close to the ear, it is observed that almost the whole energy is contained in windows of similar short duration. Therefore, a simple non-coherent receiver structures such as the energy detector with a fixed short integration duration can capture almost the whole energy in the ear-to-ear channel. The above mentioned results concerning the UWB BAN channel are presented in Part II. They were also partially published in:

- Thomas Zasowski, Frank Althaus, Mathias Stäger, Armin Wittneben, and Gerhard Tröster. UWB for noninvasive wireless body area networks: Channel measurements and results. *IEEE Conference on Ultra Wideband Systems and Technologies, UWBST*, pages 285–289, November 2003.

- Thomas Zasowski, Gabriel Meyer, Frank Althaus, and Armin Wittneben. Propagation effects in UWB body area networks. *IEEE International Conference on Ultra-Wideband, ICU 2005*, pages 16–21, September 2005.

- Thomas Zasowski, Gabriel Meyer, Frank Althaus, and Armin Wittneben. UWB signal propagation at the human head. *IEEE Transactions on Microwave Theory and Techniques*, 54:1846–1857, April 2006.

Due to their large bandwidth UWB systems are often claimed to be robust against narrowband interference [5], [13], [14]. However, this has not been shown, yet, and therefore we considered the impact of interferers on a UWB systems. In Part III, interference measurements are presented. After verification of different wireless standards by means of measurements it is shown that background noise, which is almost always present, such as GSM basestations, can be handled by filtering of out-of-band noise in the receiver. However, such a filtering is not sufficient for interferers in close vicinity of the UWB device so that a different approach has to be considered. In Part V, we propose the use of a temporal detect-and-avoid scheme, which we refer to as the temporal cognitive UWB medium access control (MAC), to avoid the interference from burst-wise transmitting wireless systems in close vicinity. Deriving an expression for the optimum packet length and data rate we show that reasonable data rates can be achieved in different interference scenarios. Moreover, packet error rates (PER) are evaluated using the proposed MAC scheme. These investigations were published in parts in:

- Thomas Zasowski, Frank Althaus, and Armin Wittneben. Temporal cognitive UWB medium access in the presence of multiple strong signal interferers. *14th IST Mobile & Wireless Communications Summit*, June 2005.

- Thomas Zasowski and Armin Wittneben. Performance of UWB Systems using a Temporal Detect-and Avoid Mechanism. *2006 IEEE International Conference on Ultra-Wideband, ICUWB*, September 2006.

The physical layer investigations are presented in Part IV. In particular, there are two noncoherent receiver structures called energy detector (ED) and transmitted reference (TR) receiver considered. In TR systems a doublet, i.e., a known reference signal and a data signal, are transmitted and correlated while the energy of the receive signal is collected by the ED. Although such receiver structures were known for quite some time [15], [16] they raised interest again for the use in UWB systems due to their simplicity compared to coherent receivers. Transmitted reference receiver for UWB was first presented in [17]. Since the reference pulse is only a noisy template, an averaging over adjacent reference signals was proposed in [18]. There, a generalized likelihood ratio test for transmitted reference systems is presented as well. A similar averaging was also presented in [19] for narrowband interference mitigation and in [20], where such an averaging has been shown to be optimal in case of no channel knowledge and not restricting to one doublet only for decision. To reduce the drawbacks on transmitting one doublet per bit, a scheme where two bits are transmitted per doublet, i.e., modulated in both amplitude and delay, was presented in [21]. However, this scheme is conceptually the same as the transmitted reference pulse interval modulation (TR PIAM) which we previously introduced in [22]. Almost at the same time a weighting after correlation in the TR receiver was proposed by [23] and [24] to improve the performance. The bit error probability for such a scheme was calculated in [25]. A closed form and upper bound for the error probability of TR systems were presented in [26] and [27], respectively. Based on an average log-likelihood ratio test a number of optimum and suboptimum receivers were obtained in [28].

To gain insights about the impact of partial channel state information (CSI) on the performance, a family of maximum-likelihood (ML) receivers with different level of CSI is derived for binary pulse position modulation (PPM), transmitted reference pulse amplitude modulation (TR PAM), and transmitted reference pulse interval modulation (TR PIAM), which is a combination of PPM and TR PAM. For the derivations, (i) full CSI, (ii) the instantaneous power delay profile (IPDP), (iii) the average power delay profile (APDP), and (iv) no CSI are assumed. As expected, the performance gets better with higher level of CSI. However, the complexity also increases with higher CSI level making the receiver structures with partial CSI less attractive for very simple WBAN applications. Since non-coherent receivers are sensitive to interference, the ML receivers in presence of inter-symbol interference (ISI) with different CSI level are also derived for binary PPM and transmitted reference binary PAM. It is shown that these receivers can handle well even strong ISI. Exemplarily, the ML receivers for transmitted reference binary PAM systems are derived in presence of a synchronous co-channel interference (CCI) and it can be observed that such receivers can cope well with the CCI. From these investigations it is concluded that the receiver for PPM with APDP knowledge is the best suited one for BAN, since it combines moderate complexity requirements with good performance results. Moreover, the integration accuracy is relaxed compared to the receiver structure without CSI. The physical layer investigations are partially published in:

- Thomas Zasowski, Frank Althaus, and Armin Wittneben. An energy efficient transmitted-reference scheme for ultra wideband communications. *International Workshop on Ultra Wideband Systems joint with Conference on Ultrawideband Systems and Technologies, Joint UWBST & IWUWBS*, pages 146–150, May 2004.

- Thomas Zasowski and Armin Wittneben. UWB transmitted reference receivers in the presence of co-channel interference. *The 17th Annual IEEE International Symposium on Personal, Indoor and Mobile Radio Communications, PIMRC*, September 2006.

- Thomas Zasowski, Florian Troesch, and Armin Wittneben. Partial Channel State Information and Intersymbol Interference in Low Complexity UWB PPM Detection. *2006 IEEE International Conference on Ultra-Wideband, ICUWB*, September 2006, (invited paper).

Finally, a recommendation for the system model is given in Part VI based on the results from the previous sections.

Chapter 3

Body Area Networks

A body area network (BAN) is a network where the communication nodes are placed directly on the body or very close to it, e.g., in the clothes [29]. This is the most general definition of a body area network. However, there exists a number of other definitions which are application oriented. Hence, these definitions are included in the aforementioned definition. One of the most frequently mentioned applications for WBANs is wireless health monitoring. There, sensors placed on the body monitor vital functions and transmit the data to a master device for evaluation [30]. Some application scenarios envision devices, that are an integral part of our everyday outfit, being always operational, context aware, and equipped to assist us in every situation [31]. Furthermore, the consumer electronics industry is also interested in WBANs in order to use it, e.g., for video gaming applications [32].

Hence, the topology of such a wearable network mainly depends on the application. The networks might have a homogeneous structure where all nodes have the same capabilities. However, for a number of applications a heterogeneous network structure is more efficient since not all nodes require the same capabilities. Such a heterogeneous BAN topology comprises many transmit-only sensor nodes, which have to be very simple, low cost, and extremely energy efficient, some transceiver nodes, which afford a somewhat higher complexity to sense and act, and few high capability nodes, e.g., master nodes with high computational capabilities, support for higher data rates, and a link to a backbone network. To allow an easy exchange of network nodes, to increase both wearing comfort and flexibility and to facilitate the communication with exposed network nodes and the environment, it is inevitable to use wireless communication in a BAN. Compared to other wireless networks one of the distinct features of a WBAN is the network topology, which is mainly determined by the shape of the human body. The shape and the size of the body limit the maximum number of nodes although network nodes can be distributed densely because of the layered clothes. Moreover, the distance between the nodes in a WBAN is very short and the transmit power shall be low due the electro-magnetic pollution. This requires a low transmit power, which should be spread over a wide frequency range to avoid peaks in the emitted spectrum. Such a system can operate in the unregulated domain below the spurious emission limits defined by the International Telecommunication Union (ITU). There exist also BAN realizations based on communication technologies such as Bluetooth working in an ISM band [33]. However, the existing wireless technologies are neither energy efficient enough nor they have a very low complexity. Moreover, Bluetooth devices are too expensive to achieve a high node density. Such devices are not optimized for the use in WBANs, neither. The capabilities of a Bluetooth device exceed the requirements of a very simple transmit only node by far, which makes the use in heterogeneous networks difficult. Hence, the existing wireless technologies are suited only to a limited extent for the use in WBANs and new communication technologies have to be found.

Chapter 4

A Brief Introduction into UWB Technologies

Starting with some historical remarks on UWB, a brief introduction into UWB technologies is presented in the following. This chapter about the basics of Ultra Wideband is an excerpt from an internal report [34], which is a joint work of Thomas Zasowski and Florian Trösch. In Section 4.2, the common definition of UWB is presented and in Section 4.3 some information on the world wide regulatory climate is given. After this, the most important promises and requirements of UWB are shown. The two common UWB realizations, i.e., impulse radio and multiband orthogonal frequency division multiplexing (OFDM), are briefly addressed in Section 4.5. An overview on the IEEE wireless personal area network (WPAN) channel models can be found in Section 4.6. Since one of the main contributions of this thesis is the derivation of new receiver structures, in Section 4.7 the basics on UWB receiver structures are presented in more detail. Finally, the current state in UWB standardization is shown in Section 4.8.

4.1 A Brief History of Ultra Wideband

Although Ultra Wideband (UWB) communication started to gain much interest in industry only few years ago it has its origin in the early days of wireless transmission. Heinrich Hertz used a spark discharge (see Fig. 4.1) that generated frequencies between 50 and 500 MHz in the 1880s and 1890s for his experiments [35]. Such spark gaps and arc discharges can be seen as first impulse radio systems and were common wave generators until the 1920s [36]. At this time, sinusoidal waveforms started to dominate in wireless communication systems. However, UWB attracted again much attention for the use in radar systems in the 1960s due to the inherent good location capabilities. Due to the wide bandwidth of UWB very accurate results in positioning could be obtained. At this time mainly the military was doing research and investigations on UWB radar. The terms "ultra wideband" and "UWB" were introduced in 1990 by the Defense Advanced Research Projects Agency (DARPA) the first time [37]. In the late 1990s, regulatory authorities in the USA started to elaborate principles for private use of UWB communications. These efforts resulted in a wide interest of UWB and universities and startup companies began to investigate UWB for private communication purposes. In 2002 as worldwide first, the Federal Communications Commission (FCC) adopted the UWB regulation for the USA. Currently, regulatory authorities in Europe and Asia are elaborating UWB regulations as well.

Figure 4.1: Original Hertzian oscillator at Institute of Physics of Polytechnical University of Karlsruhe [38]

4.2 Definition of Ultra Wideband

In general, a UWB system is considered as a system, whose bandwidth is in the same order of magnitude as its center frequency. However, since the definition of UWB systems is strongly related to the regulation in each country or on each continent there does not exist only one definition of UWB. Hence, we present in the following the first and the most common UWB definitions.

The first definition of UWB systems was given by the DARPA. According to [37], UWB systems are defined as systems that either use a bandwidth

$$B = f_u - f_l > 1500 \text{ MHz} \tag{4.1}$$

or have a fractional bandwidth

$$BW = 2 \cdot \frac{f_u - f_l}{f_u + f_l} > 0.25 \tag{4.2}$$

where f_l denotes the lower cut-off frequency and f_u the upper cut-off frequency. Both cut-off frequencies are defined at the -10 dB emission points.

In 2002, the Federal Communications Commission (FCC) adopted a UWB regulation and gave a new definition of UWB systems with less stringent requirements than the DARPA definition. This UWB definition is currently the most widely accepted definition of UWB systems. According to the FCC, UWB systems have either to occupy a bandwidth

$$B = f_u - f_l > 500 \text{ MHz} \tag{4.3}$$

or to use a fractional bandwidth

$$BW = 2 \cdot \frac{f_u - f_l}{f_u + f_l} > 0.2. \tag{4.4}$$

This UWB definition leads to two different UWB regimes. For systems with a center frequency $f_c > 2500$ MHz the requirement on the fractional bandwidth is always fulfilled and therefore the minimum bandwidth is the stringent constraint. The stringent constraint for systems with a center frequency $f_c < 2500$ MHz is the fulfillment of the fractional bandwidth constraint.

31

4.3 Regulatory Issues

Since the idea of using UWB for personal communications is relatively new, the legal use of UWB has been allowed in only a few countries up to now. In the following, we give an overview on existing UWB regulations and the currently considered UWB proposals under investigation. The average power spectral density masks adopted by different regulatory authorities are shown in Fig. 4.2.

4.3.1 USA

As the first regulatory authority worldwide, in 2002, the FCC approved UWB communications for indoor systems and handheld devices such as a laptop computer or a PDA [39] in the frequency range between 3.1 and 10.6 GHz. According to the FCC it is allowed to use UWB either indoor or in handheld devices. As shown in Section 4.2, UWB systems must have either a minimum bandwidth $B = 500$ MHz or a fractional bandwidth $BW > 0.2$. The maximum allowed transmit power spectral density is given in terms of the equivalent isotropically radiated power (EIRP). The EIRP is the product of the power supplied to the antenna and the antenna gain in a given direction relative to an isotropic antenna. In the frequency range between 3.1 and 10.6 GHz the limit for the average EIRP spectral density is $-41.3 \frac{\text{dBm}}{\text{MHz}}$, i.e., 75 nW/MHz. Thus, the maximum average EIRP of a UWB system using the full bandwidth is given by $\text{EIRP}_{\text{max}} = 0.5625$ mW. Besides the average power spectral density the FCC defined also a peak EIRP spectral density that is $0 \frac{\text{dBm}}{\text{50MHz}}$.

4.3.2 Europe

The idea of having one regulation worldwide was one of the big benefits claimed for UWB. However, the Electronic Communications Committee (ECC) decided on a more stringent regulation [40]. According to the ECC the regulation done by the FCC is not sufficient to protect the existing wireless services from UWB interference. Hence, the ECC allows UWB transmission only in the frequency band $6 - 8.5$ GHz with a maximum EIRP spectral density of

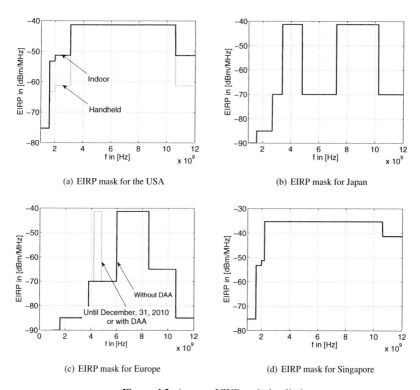

(a) EIRP mask for the USA

(b) EIRP mask for Japan

(c) EIRP mask for Europe

(d) EIRP mask for Singapore

Figure 4.2: Average UWB emission limits

$-41.3 \frac{\text{dBm}}{\text{MHz}}$. The ECC defined as well as the FCC a peak EIRP spectral density of $0 \frac{\text{dBm}}{50\text{MHz}}$. Using a phased approach, the same maximum EIRP spectral densities are allowed for the frequency range from $4.2 - 4.8$ GHz. However, the usage of this frequency range without appropriate mitigation techniques, such as a detect-and-avoid (DAA) scheme, is only allowed until December, 31, 2010 [41]. After this date, an appropriate mitigation technique is necessary for devices that operate in the frequency band from $4.2 - 4.8$ GHz. While no requirement is given for the fractional bandwidth, the minimum bandwidth for UWB systems has to be wider than 50 MHz.

4.3.3 Asia

Up to now, there exists no uniform regulation in Asia and only regulations for Singapore and Japan are accessible.

Singapore

The least stringent UWB regulation is adopted in Singapore for trial purposes. There, on a campus a UWB friendly zone was established in 2003 [42] for UWB compatibility tests. The frequency band allowed for UWB communications ranges from 2.2 to 10.6 GHz. In this frequency range the maximum average EIRP is -35.3 dBm/MHz.

Japan

In Japan, a UWB regulation is expected to be ready in 2007 [43]. To protect the ISM bands and other existing wireless services at 5 GHz, the frequency range between 4.8 and 7.25 GHz is forbidden for the use of UWB. UWB transmission is allowed only in the frequency ranges between 3.4-4.8 GHz and 7.25-10.25 GHz. In both bands the maximum average EIRP is -41.3 dBm/MHz and the peak power EIRP is 0 dBm/50MHz but in the lower band a DAA mechanism is required to mitigate interference to existing wireless systems. In the frequency range from 4.2 to 4.8 GHz such a DAA is not required until the end of 2008. In contrast to the other regulations a minimum data rate of 50 Mbit/s is required. A lower data rate is only permitted in the case where the purpose of using the lower data rate is for interference avoidance from noise and noise-like.

4.4 Promises and Challenges

Due to the above mentioned regulations, UWB systems are completely different from narrow- and wideband systems. The bandwidth of UWB systems is much larger and the maximum power spectral density much lower compared to narrow- and wideband systems as conceptually shown in Fig. 4.3. Due to this wide bandwidth UWB systems exhibit an inherently high frequency diversity [44] and are robust against narrowband interference as long as the interference power is not large enough to jam the UWB receiver [4]. Besides the interference robustness,

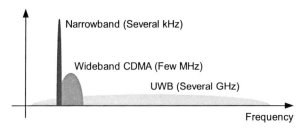

Figure 4.3: Schematic comparison between narrowband, wideband, and ultra wideband systems

the very wide bandwidth of UWB leads to a high achievable data rate. Considering Shannon's capacity formula [45]

$$C = B \cdot \log_2 (1 + \text{SNR}) \tag{4.5}$$

where C denotes the channel capacity in bits/Hz and SNR the signal-to-noise ratio it can be observed that the capacity increases linearly with the bandwidth B but only logarithmical with the SNR. In Fig. 4.4, the capacity for a UWB system with bandwidth $B = 7.5$ GHz transmitting with an EIRP of -41.3 dBm/MHz is shown as well as the capacities for the 2.4 and 5.2 GHz industrial, scientific, and medical (ISM) bands. For the 2.4 GHz ISM band a maximum transmit power of $P_{\text{TX}} = 100$ mW and a bandwidth of $B = 83.5$ MHz are assumed, and for the 5.2 GHz ISM band $P_{\text{TX}} = 200$ mW and $B = 200$ MHz. The SNR is calculated by assuming the noise power spectral density $N_0 = -174$ dBm/Hz and free space propagation between transmitter and receiver. Hence, the relation between receive and transmit power is given by the Friis' formula for an isotropic antenna [46] as

$$P_{\text{RX}} = P_{\text{TX}} \cdot \left(\frac{\lambda}{4\pi d} \right)^2 \tag{4.6}$$

where λ denotes the wavelength and r the distance between transmitter and receiver. It can be observed from Fig. 4.4 that UWB outperforms both ISM bands in terms of capacity in the

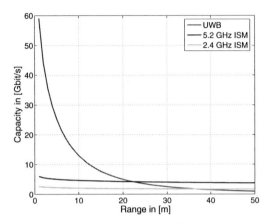

Figure 4.4: Capacity for a UWB system with $B = 7.5$ GHz compared to the 2.4 and 5.2 GHz ISM bands

short range domain. For larger distances between transmitter and receiver the UWB capacity decreases and even gets worse than the capacity that can be achieved in the ISM bands. Thus, UWB systems are a promising candidate for short range high data rate wireless communication networks. Recently, UWB also attracted the interest for use in low data rate networks. On the first glance this seems to be a waste of resources. However, due to the good scalability of UWB systems it is possible to trade data rate for transmit range. This can be done for example in an impulse radio (IR) system, which is one way to realize a UWB system by transmitting very short pulses with high bandwidth in time domain, by varying the number of successive transmit pulses per symbol. Increasing the number of transmit pulses per symbol, the data rate decreases but the receive SNR increases. These easy trade-offs of data rate and transmission range make UWB very attractive for a large number of applications, especially for such applications requiring different data rates in a heterogenous network. By using UWB, only one physical layer technology is needed to fulfill different data rate requirements. In Fig. 4.5, the envisioned region for data rate and transmission range of UWB systems is compared to a number of other wireless technologies.

Pulses in IR UWB systems have a duration of less than a few nanoseconds due to the wide

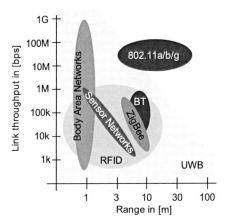

Figure 4.5: Target data rates and transmission ranges of UWB compared to existing wireless technologies

bandwidth. This fine resolution in time domain allows for localization accuracies of few centimeters that can be achieved in line-of-sight (LOS) environments [47]. Thus, besides the use of UWB for communication purposes IR UWB systems are very well suited for positioning applications. It should be noted that communication and localization are possible at the same time and that they do not mutually exclude.

Since pulse based transmission in IR systems is often considered as carrierless, many researchers expect that no mixers, radio frequency (RF) oscillators, and phased-locked-loops (PLL) are required in IR transmitters and receivers, thus resulting in a hardware of very low complexity [3]. This is probably only the case for non-coherent receivers realized in analogue domain and for the transmitter. A receiver realization in digital domain would require sampling rates in the order of some GHz, which makes such a realization impracticable for low cost UWB systems. Due to the very wide bandwidth the UWB channel is rich of multipath components that make a channel estimation very difficult. Thus, coherent receivers are currently also impracticable for UWB systems. Moreover, the very low transmit power and the short pulse duration in IR systems make the synchronization difficult. Another challenge is the development of small UWB antennas. Since narrowband antennas are usually tuned to the system's resonant frequency and are thus only working at this frequency, different antenna concepts have to be considered for developing UWB antennas.

4.5 UWB Realizations

Due to the rather generic definitions of UWB by the regulatory authorities, there exists a vast variety of possible realizations for UWB communication systems. In the following, the two most popular UWB approaches are presented.

4.5.1 Impulse Radio

The very wide bandwidth of UWB signals in frequency domain directly translates into very short pulses of several hundred picoseconds up to a few nanoseconds duration in time domain. These pulses are often considered as baseband pulses because the carrier cannot be observed from the pulse form in case that bandwidth is large compared to the center frequency. However, impulse radio pulses are bandpass pulses due to the UWB bandwidth constraints. By transmission of a pulse sequence an impulse radio system is realized [48]. The information is in most cases modulated either in the position or in the amplitude of the pulses using a pulse position modulation (PPM) or a pulse amplitude modulation (PAM), respectively. Although collisions between pulses from different UWB systems are rather unlikely if only a limited number of pulses is transmitted per second, i.e. in low duty cycle systems, pulse collisions may occur in high data rate UWB systems where the pulse repetition frequency is large. To reduce the effect of pulse collisions, different users are separated by either direct sequence (DS) or time-hopping (TH). In both cases, one symbol consists of several pulses. However, in DS UWB the pulses are multiplied by a pseudo-random sequence, while in TH UWB the pulses are mapped to different orthogonal time slots according to a user specific code sequence.

4.5.2 Multi Band OFDM

Different from the impulse radio approach, for multiband OFDM the UWB band is divided into fourteen sub-bands of 528MHz, whereby only a single band is used for OFDM transmission at a time [49]. One sub-band contains 128 OFDM carriers. The sub-band are classified in 5 band groups, which are managed according to a frequency division multiple access (FDMA) protocol. However, only one band group is mandatory for all devices while the others are optional. By application of different time-frequency codes (TFC) the sub-bands in each band group are used to establish virtual channels similar to the frequency hopping system in Bluetooth. The TFC reduce the number of collisions between different users and hence allow for a larger number of users within the same band group. Additionally, performance gains can be achieved with TFC due to the increased frequency diversity compared to a static FDMA systems, where each user is assigned to one fixed band for its overall transmission.

The main advantage of MB-OFDM systems is the flexibility in spectrum use. The spectrum can easily be adapted to different spectral masks and narrowband interferers can be avoided by omitting the affected sub-band. Although MB-OFDM systems require significantly more complex and expensive hardware designs and consume much more power than a typical IR UWB system, most companies prefer MB-OFDM because existing OFDM solutions can be transferred to the larger bandwidths. However, for applications requiring very low power consumption such as wireless sensor networks and body area networks, the MB-OFDM approach is not suited due to the power consumption.

4.6 UWB Channel Models

The most important difference between UWB and narrowband or conventional broadband systems is its extremely wide system bandwidth. Due to this ultra wide bandwidth, the UWB communication channel is strongly frequency selective and exhibits a very high multipath resolution, which leads to fundamentally different channel properties. Hence, existing narrowband channel models are not appropriate and new ones are required. Since adequate channel models are crucial in the design process of new communication systems, there has been a huge effort in UWB channel measurements and modeling aspects, which is still ongoing. This includes the characterization of main channel parameters such as path loss, fading statistics and correlation, number of multi-path components, and delay spread for different environments. The two most popular UWB channel models which were proposed by the IEEE standardization groups 802.15.3a and 802.15.4a are presented briefly in the following.

4.6.1 IEEE 802.15.3a Channel Model

The IEEE 802.15.3a channel model [50] consists of two parts, a path loss and a multipath model. The path loss is modeled as narrowband free-space path loss at the center frequency of the considered UWB band. Hence, frequency dependency, multipath effects as well the effect of LOS blockage in non-line-of-sight (NLOS) scenarios are neglected, which is an important drawback of this model [51].

The model of the multipaths is based on the Saleh-Valenzuela channel model proposed in [52]. However, the IEEE 802.15.3a model is a real bandpass model while the Saleh-Valenzuela model is formulated in the complex baseband. It is assumed in the IEEE 802.15.3a channel model that multipath components caused by the same obstacle arrive in groups, the so called clusters. Hence, the overall multipath model consists of a real continuous time impulse response according to

$$h(t) = X \sum_{l=0}^{L-1} \sum_{j=0}^{J-1} \alpha_{l,j} \delta \left(t - T_l - \tau_{l,j} \right), \tag{4.7}$$

where T_l equals the arrival time of the l-th cluster, and $\tau_{l,j}$ is the delay of the j-th multipath component in the l-th cluster with respect to the cluster arrival time T_l. The amplitude of the j-th multipath within the l-th cluster is described by $\alpha_{l,j}$ and X describes log-normal shadowing. Both cluster and ray arrival times are modeled as Poisson processes and, hence, the arrival times are according to exponential distributions. These channel parameters are defined for four different scenarios CM1-CM4. CM1 and CM2 model the channel for a distance between transmitter and receiver up to 4m in a LOS and a NLOS environment, respectively. While CM3 models a NLOS with distance between 4m and 10m, CM4 models a strong multipath scenario. Since the model neglects frequency dispersive effects during transmission, it is suited only for scenarios, where transmitter, receiver as well as environment are static or move slowly. Further information on the IEEE 802.15.3a channel model can be found in [50] and [34].

4.6.2 IEEE 802.15.4a Channel Model

The IEEE 802.15.4a channel model [53] considers three different frequency ranges, among them also the for UWB interesting range from 2 to 10GHz. For this frequency range a set of channel models is proposed for different environments such as indoor residential, indoor office, industrial environment, body-area network, outdoor, and agricultural areas or farms. The structure of the channel model remains the same for all environments, except from the body area network scenario, and only the model parameters change. As well as the IEEE 802.15.3a channel model the IEEE 802.15.4a channel model is based on the Saleh-Valenzuela model. However, there exist several differences. In the IEEE 802.15.4a channel model, channels are modeled in the complex baseband. Moreover, a frequency dependent path gain, Nakagami-m distribution of the multipath rays within a cluster, other parameter sets and a description of the cluster decay times as a function of the channel delay are introduced. Based on simulations with 2 GHz bandwidth the body-area network channel model is developed for the frequency range between 2 and 6 GHz and consists of a discrete tapped delay line model with correlated Gaussian taps scaled according to the path loss. It is assumed for this model that the transmit antenna is always mounted at the chest while the position of the receive antenna varies. The path loss is modeled by an exponential decay of power versus the distance between the antennas on the body surface. The amplitudes of the multipath components are modeled by a log-normal distribution. Moreover, a correlation of the multipaths is assumed. The number of clusters is always 2 and the inter-cluster arrival time between the two clusters is a deterministic function of transmitter and receiver location on the body. The inter-cluster as well as the inter-ray arrival times are modeled as fixed depending only on the scenario. The overall discrete stochastic BAN model is best generated in the logarithmic scale according to

$$\vec{h}\,[\,\mathrm{dB}] \;\;=\;\; \vec{x}\mathbf{B} - \vec{m} + G_{\,\mathrm{dB}}, \tag{4.8}$$

where $\vec{x}$ is a vector of N uncorrelated, unit-mean, unit-variance normal variables. The appropriate average power delay profile as well as the path correlations are introduced by an upper triangular Cholesky factorization $\mathbf{B}$ of the desired covariance matrix $\mathbf{C}$. Mean values and path loss are introduced by $\vec{m}$ and $G_{\,\mathrm{dB.}}$

4.7 Impulse Radio RX

In the following, an introduction into IR receiver structures is presented. For this overview, we classify two groups of receiver structures, coherent and non-coherent receivers. While coherent receivers can use some amount of channel state information (CSI) that allows for coherent combining, non-coherent receivers do not have any prior channel knowledge. Hence, the performance of coherent receivers is usually better, i.e., the bit error ratio (BER) is lower, than that for the non-coherent receivers but on the other hand they are essentially more complex since channel estimation has to be performed.

4.7.1 Coherent Receivers

Coherent receivers use either full or partial channel knowledge in order to reverse the influence of the channel on the transmit signal. The optimum coherent receiver is the matched filter (MF) whose derivation is given in the following. After presentation of the MF receiver the All RAKE receiver is introduced, which is one possible realization for a MF receiver. Based on the All RAKE receiver a number of less complex RAKE receivers is presented, which have a suboptimal performance. For simplicity reasons we do not consider any time-hopping in this section. However, time hopping can be easily applied to the presented receiver structures.

Matched Filter

For derivation of the MF receiver we assume the system model given in Fig. 4.6. We consider a single pulse $p(t)$ as transmit signal $s(t)$, i.e., $s(t) = p(t)$. The receive signal $r(t)$ is the sum of a single transmitted pulse $s(t)$ of duration T and zero mean white noise $n(t)$ with a power spectral density $N_0/2$. The receiver consists of a linear time-invariant (LTI) filter $g(t)$ and a sampler.

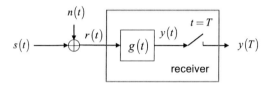

Figure 4.6: System model for derivation of the MF receiver

The receive signal $r(t)$ is given by

$$r(t) = s(t) + n(t) \tag{4.9}$$

and the signal $y(t)$ behind the filter is given by

$$
\begin{aligned}
y(t) =& s'(t) + n'(t) \\
=& g(t) * s(t) + g(t) * n(t)
\end{aligned}
\tag{4.10}
$$

where $*$ denotes the convolution operator. The MF is the filter that maximizes the SNR [54]

$$\eta = \frac{|s'(T)|^2}{\mathcal{E}\{n'^2(t)\}} \tag{4.11}$$

at the sampling time $t = T$. $\mathcal{E}$ denotes the expectation operator. Using the inverse Fourier transform

$$s'(t) = g(t) * s(t) = \int_{-\infty}^{\infty} G(f) \cdot S(f) \cdot \exp(j2\pi f t) df \tag{4.12}$$

the SNR can be written as

$$\eta = \frac{\left| \int_{-\infty}^{\infty} G(f) \cdot S(f) \cdot \exp(j2\pi f t) df \right|^2}{\frac{N_0}{2} \int_{-\infty}^{\infty} |G(f)|^2 \, df}. \tag{4.13}$$

Applying the Schwarz inequality

$$\left| \int_{-\infty}^{\infty} a(x) \cdot b(x) dx \right|^2 \leq \int_{-\infty}^{\infty} |a(x)|^2 \, dx \cdot \int_{-\infty}^{\infty} |b(x)|^2 \, dx \tag{4.14}$$

to (4.13) we get for the SNR

$$\eta = \frac{\left| \int_{-\infty}^{\infty} G(f) \cdot S(f) \cdot \exp(j2\pi ft) df \right|^2}{\frac{N_0}{2} \int_{-\infty}^{\infty} |G(f)|^2 \, df} \leq \frac{\int_{-\infty}^{\infty} |G(f)|^2 \, df \cdot \int_{-\infty}^{\infty} |S(f)|^2 \, df}{\frac{N_0}{2} \int_{-\infty}^{\infty} |G(f)|^2 \, df} = \frac{2E_s}{N_0}$$

$$\tag{4.15}$$

where E_s denotes the signal energy. Equality is given in (4.15) for a filter with the transfer function

$$G_{opt}(f) = k \cdot S^*(f) \cdot \exp(-j2\pi fT) \tag{4.16}$$

where $S^*(f)$ is the conjugate complex of $S(f)$ and k is an arbitrary scaling factor. The time domain representation of the matched filter is given by

$$g_{opt}(t) = k \cdot s(T - t). \tag{4.17}$$

Thus, for a single path channel the matched filter is a time reversed, delayed, and with k scaled version of the transmit signal $s(t)$, i.e., a signal matched filter.

As shown in Section 4.6 the energy of UWB–IR indoor channels is spread over a large number of multipath components. For multipath channels, the signal matched filter is not anymore optimum since the channel impact is not considered there. The system model of the matched filter for multipath channels with the channel impulse response (CIR) $h(t)$ is shown in Fig. 4.8. The receive signal is here given by

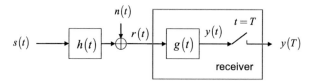

Figure 4.7: System model of the matched filter for multipath channels

$$r(t) = h(t) * s(t) + n(t).$$ (4.18)

Hence, the filter $g(t)$ has now not only to be matched to the signal but also to the channel. This channel matched filter can be written as

$$g(t) = k \cdot (h^*(t - T) * s^*(t - T)).$$ (4.19)

All RAKE

Another way to realize an optimum receiver is the correlation receiver. The system model of the correlation receiver for a multipath channel is shown in Fig. 4.8. The receive signal $r(t)$ is

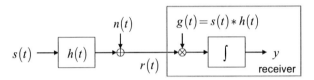

Figure 4.8: System model of the correlation receiver for multipath channels

correlated with

$$g(t) = s(t) * h(t)$$ (4.20)

yielding the decision variable

$$y = \int r(t) \cdot g(t) dt.$$ (4.21)

Assuming a tap delay line channel, $h(t)$ can be represented by

$$h(t) = \sum_{i=1}^{N} h_i \cdot \delta(t - \tau_i)$$ (4.22)

where N denotes the number of channel taps, h_i the amplitude, and τ_i the delay of the ith channel tap. With (4.22) the result in (4.20) can be written as

$$g(t) = \sum_{i=1}^{N} h_i \cdot s(t - \tau_i).$$ (4.23)

Thus, the signal $g(t)$ which is used for correlation with the receive signal $r(t)$ is represented by a sum of delayed transmit signals $s(t - \tau_i)$ weighted with the corresponding channel taps h_i. Using this representation of $g(t)$ and changing the order of summation and integration, the correlation receiver structure in Fig. 4.8 can be modified and results in the receiver structure shown in Fig. 4.9. This specific realization of an optimum receiver is known as the All RAKE (ARAKE), which was presented for the first time in [55]. In each branch of the All RAKE, the signal portion caused by one channel tap is correlated with a delayed version of the transmit signal $s(t - \tau_i)$ and weighted with the amplitude of the corresponding channel tap h_i. Thus, the number of branches has to equal the number of channel taps. The results of all branches are summed up and yield the decision variable

$$y = \sum_{i=1}^{N} h_i \cdot \int r(t) \cdot s(t - \tau_i) dt. \tag{4.24}$$

Frequently, the correlator branches are called RAKE fingers.

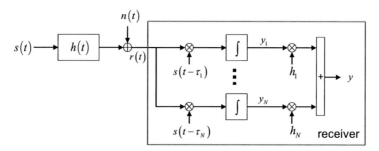

Figure 4.9: System model of the All RAKE receiver for multipath channels

Selective and Partial RAKE

Since the number of channel taps N in UWB channels is usually very high an All RAKE would require a very high number of correlators. Besides this, it can be seen from Fig. 4.9 that the receiver must have full knowledge of all channel tap amplitudes h_i and all delays τ_i. This requires an accurate channel estimation that is challenging for UWB systems. Thus, the implementation of the All RAKE in UWB systems is currently impracticable. However, the complexity of an All RAKE receiver can be reduced if the channel taps are separable, i.e., if the delayed pulses do no overlap at the receiver. In such a case, $g(t)$ in (4.23) can be approximated by using only a subset of all present channel taps. Two suboptimum RAKE receiver structures are the Selective RAKE (SRAKE) and the Partial RAKE (PRAKE) [56]. The SRAKE has only L, $(L < N)$, fingers where the receive signal contributions of the L strongest paths are correlated. Therefore, the number of correlators required in the receiver can be reduced. However, the knowledge of the instantaneous values of all multipath components is needed requiring a full channel estimation. The PRAKE also has a reduced number of RAKE fingers L, $(L < N)$. In contrast to the SRAKE, the receive signal components caused by the L first paths are correlated in the PRAKE. Using a PRAKE, only knowledge of the L first paths is required resulting in a less complex channel estimation. Since the first paths are not necessarily the strongest, especially in non line-of-sight (NLOS) channels, the SRAKE usually exhibits a better performance than the PRAKE with the same number of RAKE fingers.

4.7.2 Non-coherent Receivers

For several applications such as body area or sensor networks even the suboptimum RAKE receiver structures are too complex. Hence, simpler receiver structures without requiring a number of correlations and a complex channel estimation would be desirable. Two non-coherent receiver structures not requiring any channel knowledge are the transmitted reference (TR) receiver and the energy detector.

Transmitted Reference Receiver

TR systems were presented for the first time in the 1960s [15]. In a TR system, two pulses, also referred to as a doublet, are transmitted for one symbol. The first pulse is a reference pulse and the second, delayed by T_d, is a data pulse, carrying the information. This transmission scheme is exemplarily depicted in Fig. 4.10. For transmitting $a = 1$, two pulses of duration T_p with positive amplitude are transmitted while for transmitting $a = -1$ one positive and one negative pulse are transmitted. Thus, the transmit signal $s(t)$ can be described by

$$s(t) = p(t) + a \cdot p(t - T_d) \tag{4.25}$$

and the receive signal $r(t)$ is given by

$$r(t) = h(t) * s(t). \tag{4.26}$$

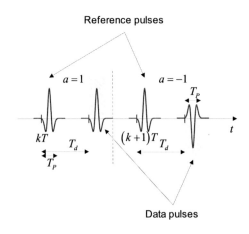

Figure 4.10: Signaling in transmitted reference systems

Assuming that the channel is time invariant over one doublet, the transmission of reference and data pulse over the same channel yields an implicit channel estimation. The reference signal is used in the receiver as template for the correlation with the data signal. It has to be noted that the reference signal is only a suboptimum, noisy template due to the additive noise. The typical transmitted reference receiver structure is shown in Fig. 4.11. In such a transmitted reference receiver the reference signal is delayed by T_d and correlated with the data signal. The decision variable y is then given by

$$y = \int_{kT+T_d}^{kT+T_d+T_p} r(t) \cdot r(t - T) dt \tag{4.27}$$

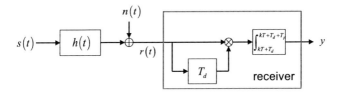

Figure 4.11: System model of the transmitted reference receiver

assuming a single tap channel. For a multipath channel the integration duration has to be adapted. The performance of a transmitted reference system strongly depends on the integration duration. If the integration time is longer than the delay spread of the channel almost noise only is collected after a certain point degrading the performance. Using a too short integration time results in missing a noticeable portion of the energy available in the receive signal.

Energy Detector

Another receiver structure not requiring any channel knowledge is the energy detector [16]. The energy detector collects, as the name implies, the energy from the multipath components of the channel. Due to the collection of energy it is not possible to use antipodal signaling but either pulse position modulation (PPM) or on-off keying (OOK). In the following an energy detector for a 2PPM is exemplarily presented. The signaling for 2PPM is shown in Fig. 4.12. One

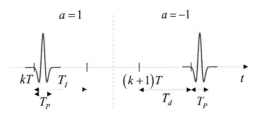

Figure 4.12: 2PPM signaling for energy detectors

possible realization for the energy detector is depicted in Fig. 4.12. The receive signal $r(t)$ is squared and the energy at the two possible time instances compared yielding the decision variable

$$y = \int_{kT}^{kT+T_p} r^2(t)dt - \int_{kT+T_d}^{kT+T_d+T_p} r^2(t)dt. \tag{4.28}$$

Depending on which window contains more energy either a $+1$ or a -1 is detected. At the time instance where no pulse is located, only noise is collected in the receiver. Similar to the transmitted reference receiver, the energy detector is very sensitive to the integration time.

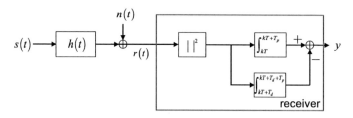

Figure 4.13: System model of the energy detector

4.8 Standardization and Trends

Since UWB communication has some very attractive properties the Institute of Electrical and Electronics Engineers (IEEE) originated two task groups working on UWB standardization in the IEEE 802.15 working group for wireless personal area networks (WPAN). The task group 3a was installed to develop a high data rate physical layer while the task group 4a has been working on a low data rate physical layer standardization.

4.8.1 IEEE 802.15.3a

The IEEE 802.15 task group 3a was established in December 2001. Goal of this task group was the development of a high data rate UWB standard. The envisioned data rates should be at least 110 Mbit/s at 10 meters distance. During the down-selection procedure, a number of proposals were rejected or merged into new proposals. At the end of the down-selection procedure one DS-UWB and one MB-OFDM proposal remained, both supported by a number of companies. Since July 2004, neither of them got the 75% level at the confirmation vote and the standardization efforts reached a point of deadlock. Thus, in January 2006 it was decided to dissolve the task group 3a. The companies now start to create de-facto standards by pushing products using their UWB solution into the market.

4.8.2 IEEE 802.15.4a

Since high data rates are for many applications less important than other requirements such as power consumption, the IEEE 802.15 task group 4a was established in March 2004. The most important requirements considered in the standard are a high aggregate throughput, ultra low power, and the location capability during the communication. The per link bit rate should be higher than 1 kbit/s while the aggregate throughput at a data collector should be at least 1 Mbit/s. For localization an accuracy of at least 1 m should be achieved having a maximum transmit range of 30 m. At this point of time a dual communication system is considered that consists of a pulse based UWB system, which supports localization, and a 2.4 GHz chirp spread spectrum system. Since the standardization process is ongoing only few detailed information is available. According to [57], for the UWB system a baseline mode is defined with a single UWB band centered at about 4 GHz is foreseen. Using a hybrid PPM+BPSK modulation a data rate of 851 kbps can be achieved. Due to the low duty cycle, collisions are very unlikely and Aloha shall be used for the medium access control. However, realizations different from the baseline mode are currently under investigation yielding different data rates.

Chapter 5

Basic System Considerations

As mentioned in Section 3, body area networks have to fulfill a number of requirements. The probably most stringent requirements are the energy efficiency, the size, and the complexity of a wireless node in a body area network. Hence, multiband OFDM UWB realizations [49] are too complex and not energy efficient enough for the use in WBANs. UWB impulse radio [58] is a very promising technology for body area networks. Due to the quasi carrierless nature of impulse radio a major reduction of the analog RF part can be achieved. No frequency synthesis, PLL, voltage controlled oscillator (VCO), mixer, and power amplifier are required [3]. Impulse radio transmitters are more or less pulse generators, only. Sometimes they exist in conjunction with a pulse shaping circuit. Using a step recovery diode (SDR) and short-circuited stubs, where the signal is reflected, is one typical method to generate short UWB pulses as shown in [59]. However, pulse generators realized in complementary metal oxide semiconductor (CMOS) technology were presented recently, e.g., in [60] and [61], since CMOS transmitters are more energy efficient than analog transmitter realizations [62]. On the receiver side simple hardware realizations are a more challenging task. In particular, coherent receivers have a high complexity because a channel estimation is required for the template estimation. Moreover, many correlations are required in a coherent receiver if the channel has a large number of multipath components [63]. Non-coherent receivers suffer from some performance degradation but require a much lower complexity [64]. There, only one correlation and no channel estimation are necessary. The RF part of transmitted reference receiver and energy detector can be implemented very efficiently in analog hardware [65]. Due to this low complexity and due to the energy efficiency, non-coherent receiver structures such as the energy detector and the transmitted-reference receiver are much more promising solutions for the use in UWB BAN. Hence, only energy detector and transmitted-reference based receiver structures are considered in the remainder.

There exist numerous WBAN applications, which require different assumptions on the node capabilities and also on their positions. Due to this reason, a number of different links at the human body is considered for the channel investigations. The head is of particular interest for the placement of wireless nodes, because most human communications organs such as mouth, ears, or eyes are located at the head. Hence, for a number of evaluations the ear-to-ear link is considered, which can be regarded as a kind of worst case scenario due to the missing line-of-sight component.

Part II

The Body Area Network Channel

Chapter 6

Channel Measurements

When developing a communication system it is inevitable to know the channel characteristics. Since the interest in BAN applications started recently, a number of investigations of the BAN channel have been performed in the meanwhile. A scenario with only one antenna mounted on the body was considered in [6]. There, the impact of the body on UWB transmission in an indoor environment was investigated in the frequency range between 1 and 11 GHz. It was shown that the body causes a notch of about 24 dB in the receive signal pattern if the body is located between both antennas. In [8], the UWB BAN channel between 3 and 6 GHz was investigated. Measurements were performed in an indoor environment to achieve realistic propagation conditions. It was shown that frequency correlation properties of the channel change substantially over the investigated frequency band. Furthermore, significant variations of signal energy spread in time-delay domain were observed. An investigation of the UWB BAN channel based on Finite Difference Time Domain (FDTD) simulation was presented in [9]. Based on these simulations in the frequency range between 2 and 6 GHz the authors derived a path loss model and evaluated the power delay profile as well as the amplitude distribution. Based on measurements from 3 to 6 GHz in a parking lot the simulation results were verified. An extension to [9] was presented in [10], where the authors noticed ground reflections that could assist communication from one side of the body to the opposite. Despite this, a strong impact of the arm motion on the receive power was observed. In [12], an evaluation of the path loss for two different antennas was shown. For these investigations the frequency range between 3 and 9 GHz was considered. A channel model for propagation around the human body between 3 and 6 GHz based on 144 measurements was presented in [11].

In the following section, the measurement setup is presented, which is used to determine the channel properties and to derive the channel model. Since the most important communication organs such as the ears are placed at the head, transmission at the head is of particular interest. Hence, channel measurements for the ear-to-ear link, which can be regarded as a worst case scenario, are shown in the successive section.

6.1 Measurement Setup

The channel measurements were performed in an anechoic chamber with several test persons and a head phantom. The head phantom is used for determination of the antenna pattern changes close to the body. Transfer functions were measured with a network analyzer in the frequency range between 2 and 8 GHz. Two meander line antennas Skycross SMT-3TO10M-A [66], which were mainly chosen due to their small size and their large bandwidth, were used for these measurements. According to the manufacturer the antennas are suited for a frequency range from 3.1 to 10 GHz and have a dipole-like antenna pattern. The antenna specifications were verified by measurements of the antenna pattern in the anechoic chamber. The corresponding antenna pattern is shown in Fig. 6.1 exemplarily for five different frequencies. From this figure it can be observed that the antenna pattern is omnidirectional. The variation of the antenna gain is relatively small over the frequency range from 3.1 to 8 GHz. Below 3.1 GHz the gain decreases as it can be seen from the antenna pattern for 2.02 GHz. Moreover, a slight directivity of the antenna patterns towards $90°$ can be observed.

To reduce the influence of unwanted cable effects on the measurements the antennas were mounted on glass-fiber reinforced plastic (GRP) arms on tripods as shown in Fig. 6.2 for a measurement with the head phantom. The tripods were covered by absorbing material to reduce reflections caused by them. With such a measurement setup only the antennas were placed close to the head while the cables led away from the head as fast as possible.

Using the above described measurement setup, transfer functions were determined for 20

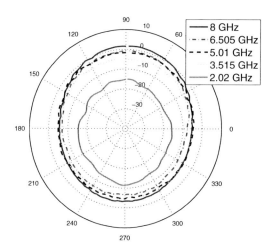

Figure 6.1: Antenna pattern of the Skycross SMT-3TO10M-A antenna measured in an anechoic chamber

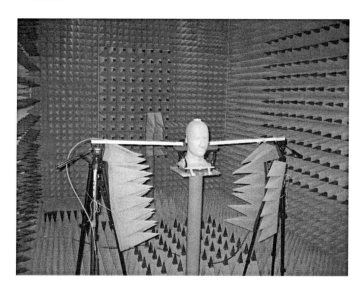

Figure 6.2: Measurement setup with the SAM head phantom in the anechoic chamber

different links at the human body. Each measurement was done 5 times on 11 different persons, resulting in 55 measurements per link, i.e., 1100 measurements altogether. To consider also the effects due to arm motions, three measurements were done with the antenna placed at the wrist, i.e., with the wrist slightly in front of the body, besides the body, and slightly behind the body. Thus, the following links were chosen for the measurements: on ear - on ear, behind ear - behind ear, behind ear - wrist in front, behind ear - wrist backward, behind ear - wrist laterally, behind ear - belt buckle, shoulder - belt buckle, shoulder - belly, shoulder - belly, shoulder - chest, shoulder - wrist in front, shoulder - wrist backward, shoulder - wrist laterally, belt buckle - belly, belt buckle - hip, belt buckle - back, belt buckle - knee, belt buckle - forefoot, belt buckle - heel.

As mentioned above, some measurements were also done with a specific anthropomorphic mannequin (SAM) head phantom V4.5 from SPEAG [67]. The phantom was filled with the head tissue simulation liquid HSL 5800, which consists of water, mineral oil, emulsifiers, additives, and salt. This lossy dielectric liquid matches the requirements according to FCC [68] in the frequency range from 4.9 to 6.0 GHz. At 5.2 GHz relative permittivity is given by $\varepsilon_r = 36.0$ and conductivity by $\sigma = 4.66\frac{S}{m}$. The values for relative permittivity and conductivity at 5.8 GHz are $\varepsilon_r = 35.3$ and by $\sigma = 5.27\frac{S}{m}$, respectively. The head phantom was placed on a plastic pillar as shown in Fig. 6.2 or on a rotating platform for measurement of the antenna patterns.

6.2 Measurements Results for the Ear-to-Ear Channel

In Fig. 6.3, frequency transfer functions for the ear-to-ear link are shown exemplarily for four different persons and the head phantom. It can be observed that the transfer functions for the test persons exhibit similar trends. The transfer function for the head phantom shows also a similar shape. However, for a wide range of frequencies the attenuation is about 5-10 dB smaller compared to the test persons. Although, the liquid inside the head phantom is only specified for the frequency range from 4.9-6 GHz, the attenuation characteristic of a human head seems to be fitted over a wider frequency range. For comparison reasons also the attenuation calculated with the Friis formula for an isotropic antenna given in (4.6) and a reference measurement

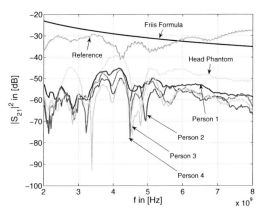

Figure 6.3: Frequency transfer functions of the ear-to-ear link for different persons and head phantom measured in an anechoic chamber

without head are shown in Fig. 6.3. The difference between the attenuation of the reference measurement and the one calculated with the Friis formula is mainly caused by the gain variations of the antennas. At the lower end of the considered frequency band we are in the Fresnel zone of the antenna. Above 3 GHz the gain of the antenna is increasing due to a constant size of the antenna but decreasing wave length. These effects can also be noticed from the antenna patterns in Fig. 6.1. For the head measurements a 20-30 dB higher attenuation can be observed compared to the reference measurement and the attenuation derived from the Friis formula. For some frequencies the transfer functions in Fig. 6.3 exhibit steep notches. Such notches can be caused by two paths that have a distance difference of $\lambda/2$, which leads to a mutual cancelation of the signals. For the two steep notches at about 3.4 GHz and 4.5 GHz there have to exist two path pairs with a distance difference of 0.044 and 0.034 m, respectively. Since there exist more than one path for the ear-to-ear link and since these distance differences are reasonable for the scenario, it can be assumed that the notches are caused by multipath propagation.

In Fig. 6.4, the impulse responses corresponding to the frequency transfer functions from Fig. 6.3 are shown for the different persons. As well as for the frequency transfer functions it can be seen that all impulses responses have a very similar shape independent of the person.

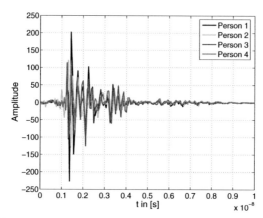

Figure 6.4: Impulse responses corresponding to the frequency transfer functions from Fig. 6.3 for the different test persons

Moreover, it can be observed that most energy is contained within 1 and 4 ns for all impulse responses. The highest peaks can be seen at the beginning of the channel impulse responses. The delay between the beginning of the measurement, i.e., 0 ns, and the first peak is mainly caused by the propagation time between transmitter and receiver.

Chapter 7

Channel Model

Besides the IEEE 802.15.4a BAN channel model presented in section 4.6, a channel model for UWB BAN is given in [11] and [69] by the same authors. The model is based on 144 frequency domain measurements between 3 and 6 GHz, where the transmitter was placed on the front side of the torso and the receiver on various positions around the torso. Hence, the authors consider three different scenarios, i.e., a receiver placed at the front, at the side, and at the back. For these three scenarios the authors determined that the amplitudes are lognormal distributed. In the following a simple channel model is derived. This model is used in section 12.7 for the performance evaluation of different receiver structures. Since the position of transmitter and receiver is usually not known for many body area network applications, different scenarios are not distinguished, i.e., the model is based on all measurements described in section 6.1.

7.1 Measurement Evaluation

The channel model shall be used for performance simulation of non-coherent receivers. Hence, a simple model for the channel impulse response amplitudes is derived in the following. All transfer functions are transformed into complex channel impulse responses by means of an inverse Fourier transform. Since the energy in the channel impulse responses at the human body decays very fast, only the first 15 ns, i.e., 240 measurement samples with a sampling rate of 16 GHz, of each impulse response are considered for the modeling. The 15 ns correspond to a distance of 4.5 m in free space. Due to the size of human bodies and since measurements were done in anechoic chamber, it can be assumed that no reflections are received after the considered time. The impulse responses are aligned such that they begin at the time instance, where the first substantial increase in energy can be observed. All channel impulse responses are normalized by the path loss for each link. Thus, the parameters of the distribution for each sample are independent of its position.

Distribution of Channel Taps   While it can be easily shown that the distribution of the phases is uniform within $\{0, 2\pi\}$, the determination of the amplitude distributions requires more effort. To determine the statistical distribution of the amplitudes, the Akaike information criterion (AIC) is used [70]. The AIC is based on the Kullback-Leibler (KL) distance, which can be used to determine the similarity of two different probability density functions. It is possible to write the KL distance as a difference between two statistical expectations [71] as

$$
\begin{aligned}
I(f,g) &= \int f(x) \log(f(x)) dx - \int f(x) \log(g(x|\Theta)) dx \\
&= \mathcal{E}_f[\log(f(x))] - \mathcal{E}_f[\log(g(x|\Theta))],
\end{aligned}
\tag{7.1}
$$

each with respect to the real distribution f, which is given. The distribution g depends on a set of parameters Θ and should approximate f. The KL distance is always larger than or equal to zero, i.e., $I(f,g) \geq 0$. The equality holds only if $f(x) = g(x|\Theta)$. The first term $\mathcal{E}_f[\log(f(x))]$ is not known, because it depends on the true distribution but it can be regarded as a constant. Hence, (7.1) can be written as

$$
I(f,g) = C - \mathcal{E}_f[\log(g(x|\Theta))]
\tag{7.2}
$$

or

$$
I(f,g) - C = -\mathcal{E}_f[\log(g(x|\Theta))].
\tag{7.3}
$$

The term $I(f,g) - C$ is regarded as a relative distance between f and g. Thus, it is sufficient to consider only $\mathcal{E}_f[\log(g(x|\Theta))]$ for calculation of a relative distance. Defining the best fitting model, $\mathcal{E}_f[\log(g(x|\Theta))]$ has to be maximized because $I(f,g) \geq 0$. It is shown in [72] that for an indefinitely large number of observations N

$$\lim_{N \to \infty} \frac{1}{N} \cdot \sum_{n=1}^{N} \log(g(x_n|\Theta)) = \int f(x) \log(g(x|\Theta)) dx$$

$$= \mathcal{E}_f[\log(g(x|\Theta))] \qquad (7.4)$$

and that maximizing $1/N \cdot \sum_{n=1}^{N} \log(g(x_n|\Theta))$ with respect to Θ yields the maximum likelihood estimate $\hat{\Theta}$, i.e.,

$$\hat{\Theta} = \arg\max_{\Theta} \left[\frac{1}{N} \cdot \sum_{n=1}^{N} \log(g(x_n|\Theta)) \right]. \qquad (7.5)$$

Using (7.5), Akaike defined its information criterion [70] as

$$AIC = -2 \cdot \frac{1}{N} \cdot \sum_{n=1}^{N} \log(g(x|\hat{\Theta})) + 2 \cdot K, \qquad (7.6)$$

where K denotes the number of estimable parameters and the AIC is regarded as an estimate for the approximation quality of different distributions. The log-likelihood $\frac{1}{N} \cdot \sum_{n=1}^{N} \log(g(x|\hat{\Theta}))$ is asymptotically biased. With the simple expression K as an estimator of $\frac{1}{N} \cdot \sum_{n=1}^{N} \log(g(x|\hat{\Theta}))$ for the asymptotic bias an approximately unbiased estimator is achieved. A more intuitive explanation of the AIC is as follows. $I(f,g)$ can be decreased by using additional known parameters for the approximation g, because g can be chosen closer to f for a fixed parameter set. Since these additional parameters are usually not known but have to be estimated, further uncertainty is added to the estimation of $I(f,g)$. Hence at a certain point, adding still more parameters has a negative effect. The Kullback-Leibler distance $I(f,g)$ will increase due to noise in the estimated parameters that are not necessary to achieve a good model. This effect is considered in the AIC. While $\frac{1}{N} \cdot \sum_{n=1}^{N} \log(g(x|\hat{\Theta}))$ decreases, K increases with additional number of parameters. Thus, models with a high number of parameters are penalized. To rank different distributions, the AIC differences

$$\Phi_j = AIC_j - AIC_{\min} \qquad (7.7)$$

are used [71]. AIC_j denotes the AIC value of the jth distribution and AIC_{min} denotes the minimum AIC indicating the distribution with the best fit. The likelihood of a model g_j with given data is computed by

$$\mathcal{L}(g_j|x) \propto e^{-\frac{1}{2}\Phi_j}. \tag{7.8}$$

The likelihood values for the different models are used to calculate the Akaike weights [73]

$$w_j = \frac{e^{-\frac{1}{2}\Phi_j}}{\sum_{i=1}^{J} e^{-\frac{1}{2}\Phi_i}} \tag{7.9}$$

which satisfy $\sum_{j=1}^{J} w_j = 1$. The weight w_j is an estimate of the relative likelihood that a distribution is the best fit to the true data within a candidate set. Hence, the Akaike weights give besides the selection of the best candidate also information on the relative approximation quality.

For determination of the amplitude distributions of the measured channel impulse responses frequently used distribution functions in UWB channel modeling [74] are used, i.e., Nakagami, Rice, Lognormal, Weibull, and Rayleigh distribution. While the first 4 distributions depend on 2 parameters, i.e., $K = 2$, the Rayleigh distribution depends on 1 parameter only. The distribution functions are given in the Appendix A.1 - A.5.

In Fig. 7.1, the Akaike weights of the different distributions are plotted. It can be observed that Rice and Nakagami distribution have only very small weights, i.e., it is very unlikely that the amplitudes are distributed according to these distributions. Although the weights for the Rayleigh and the Weibull distribution are slightly larger, they are also small compared to the Akaike weights for the lognormal distribution. Therefore, it is assumed that the amplitudes are lognormal distributed according to

$$f(x) = \frac{1}{\sigma_x \sqrt{2\pi}} e^{-\frac{(\ln(x)-\mu)^2}{2\sigma^2}}, \tag{7.10}$$

since these Akaike weights are in general the highest, i.e., the lognormal distribution is the most likely one.

To verify the assumption that the channel taps are lognormal distributed the cumulative distribution functions (cdf) of the measurements are compared to the theoretical ones with parameters

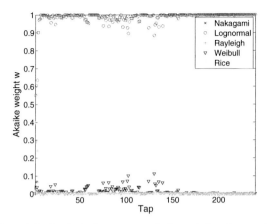

Figure 7.1: Akaike weights for the body area network channel plotted versus the corresponding channel tap

determined by a maximum likelihood estimation. In Fig. 7.2, the cdfs are shown for one exemplarily chosen channel tap. It can be seen that neither the cdfs for Rayleigh, Rice, Weibull nor Nakagami distribution fit the measured distribution well. However, the lognormal distribution fits the measured cdf very well over the whole range of values. Thus, the result of the AIC method is verified. This observation corresponds also with the results in [11].

The parameters μ and σ for the lognormal distribution in (7.10) determined by a maximum likelihood estimation and averaged over all normalized channel taps are $\mu = -0.60$ and $\sigma = 0.84$.

Power Delay Profile and Path Loss   Besides the distribution of the channel taps also the power delay profile and the path loss are of interest. The power delay profile (PDP), averaged over all channel measurements, is shown in Fig. 7.3. A decay over the time can be observed from this plot. Since the PDP does not show a linear decay in the logarithmic domain over the considered dynamic range, a function that approximates the power delay profile has to be defined for the model. From Fig. 7.3 two ranges with different linear decays can be observed. Up to about 2 ns the PDP is decaying relatively steep. For delays above about 2 ns the linear

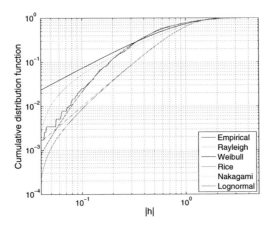

Figure 7.2: Cumulative distribution functions for an exemplarily chosen channel tap

decay gets flatter. This flattening is caused by effects such as reflections from the body or also from the environment. Although the measurements were done in an anechoic chamber, in particular, the absorbing material placed on the tripods is only attenuating the waves by about 15 dB and can still cause strongly attenuated reflections.

Due to this behavior of the PDP in logarithmic scale the PDP shown in Fig. 7.3 is approximated by two exponentials for the different ranges. The approximation for the range up to 2 ns is given by

$$A = a_1 - a_2 \cdot t \tag{7.11}$$

and for the range above 2 ns by

$$B = b_1 - b_2 \cdot t, \tag{7.12}$$

where t denotes the time in seconds. The parameters a_1, a_2, b_1, b_2 are determined as $a_1 = -0.8$, $a_2 = 12.5 \cdot 10^{-9}$, $b_1 = -21.0$, and $b_2 = 1.9 \cdot 10^{-9}$ using a least square curve fitting in the logarithmic domain.

From the APDP it can be concluded that the main energy is contained in a very short time interval. Only about 2 ns after the maximum peak the energy is decayed by about 40 dB.

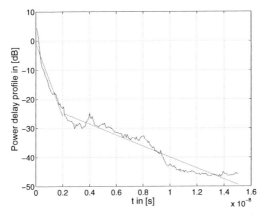

Figure 7.3: Power delay profile over all channel measurements and approximation of the APDP based on (7.11) and (7.12)

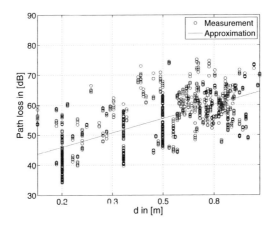

Figure 7.4: Measured and approximated path loss for the body area network channel

The path loss (PL) can be calculated directly from the measured frequency transfer functions [75]. If there are M transfer functions available for a distance d with N frequency points, the average path loss is given by

$$\text{Pl}(d) = \frac{1}{MN} \cdot \sum_{i=1}^{N} \sum_{j=1}^{M} \left| H_j^{(d)}(f_i) \right|^2 . \tag{7.13}$$

$H_j^{(d)}(f_i)$ denotes the jth frequency transfer function at a frequency f_i in a distance d. The distance d is the distance between transmitter and receiver on the surface of the body. Assuming $\text{PL}(d) \propto d^{\gamma}$ the path loss exponent γ can be evaluated at any distance d as described in [76] by

$$\text{PL}(d) = \text{PL}_0 + 10 \cdot \gamma \cdot \log_{10} \left(\frac{d}{d_0} \right) + S_\sigma \tag{7.14}$$

where PL_0 denotes the path loss at a distance d_0. d_0 is set to $d_0 = 0.1$ m and S_σ is a lognormal variable with standard deviation σ which accounts for the path loss variations at a fixed distance. To determine the path loss $\text{PL}(d)$, which includes in this case attenuation, reflection, and diffraction effects, a least square fit computation is performed yielding a reference path loss

$$\text{PL}_0 = 38.9 \text{ dB} \tag{7.15}$$

and a path loss exponent

$$\gamma = 2.4. \tag{7.16}$$

The standard deviation σ of the lognormal distributed variable S_σ is determined as

$$\sigma = 6.8 \text{ dB} \tag{7.17}$$

using a maximum likelihood estimation. In Fig. 7.4, the measured path losses $\text{Pl}(d)$ are displayed as blue circles for the different distances d, while the approximation $\text{PL}(d)$ is shown as magenta line. It can be observed that the path loss is varying over a wider range for similar distances d. These variations are mainly caused by the fact that there exist several measurements with similar distances for different links and different persons.

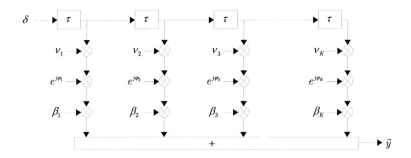

Figure 7.5: Tap delay line representation of the channel model

7.2 Generation of an Impulse Response

Using the above determined parameters channel impulse responses can be generated according to the tap delay line representation of the channel model in Fig. 7.5. For each tap of the impulse response the impulse δ is delayed by $\tau = 62.5$ ps. In each branch the impulse is multiplied by the lognormal magnitude $\nu_i \sim \mathcal{LN}\{-0.60, 0.84\}$ and the exponent $e^{j\phi_i}$ with phase ϕ_i equally distributed in $\{0, 2\pi\}$. According to the delay of the channel tap, afterwards a multiplication with the corresponding linear value β_i of the power delay profile calculated from (7.11) and (7.12) is done. Finally, all taps are summed up and multiplied with the distance dependent path loss $\text{PL}(d)$ given in (7.14) yielding the modeled channel impulse response vector $\vec{y}$.

Exemplarily, a measured and a modeled channel impulse response are shown in Fig. 7.6. It can be seen that both impulse responses are relatively similar. After a strong peak at the beginning both impulse responses decay very fast.

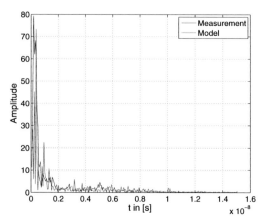

Figure 7.6: Exemplary absolute values of the measured and modeled complex channel impulse responses

Chapter 8

Impact of the Channel Properties on the System Design

In this section, the impact of the channel on the system design and the antenna placement is investigated. For this investigation the ear-to-ear link is considered again. By means of measurements, theory, and simulation it is shown that the direct signal path through the head is negligible. After discussing the implication of this effect on the antenna pattern different distances between antenna and skin are considered in the measurements. From these measurements it can be observed that the channel is robust against such distance variations. Comparing antenna patterns with and without head it can be seen that the proximity of antenna and skin improves the antenna's radiation characteristics such that reflections away from the head are present. Based on measurements, where the antennas were placed on the ears, behind the ears, above the ears, and in front of the ears, the impact of different antenna positions for the ear-to-ear link is discussed. Finally, some conclusions are drawn on how the energy distribution of the channel impulse responses influences the system design. These investigations were partially published in [77] and [78].

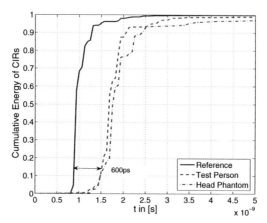

Figure 8.1: Cumulative energy of channel impulse responses for transmission with and without
head

8.1 Direct Transmission Negligible

Wavelength and propagation speed of electromagnetic waves depend on the material parameters
of the propagation medium. Thus, the expected difference in the arrival time of a direct path
through the head and the reference path through the air can be calculated according to (8.1).
Based on the permittivity and conductivity values given for the head phantom a direct path
through the head would have an expected delay of

$$\Delta t = \frac{d}{v_{\text{head}}} - \frac{d}{c} = 2.92\text{ns} \tag{8.1}$$

compared to the direct path of a reference measurement without head. In (8.1), $v_{\text{head}} = 4.88 \cdot 10^7 \frac{\text{m}}{\text{s}}$ denotes the propagation speed through the head, $c = 3 \cdot 10^8 \frac{\text{m}}{\text{s}}$ the light speed, and
$d = 0.17$ m the distance between both antennas.

In Fig. 8.1, cumulative energies of channel impulse responses measured in an anechoic chamber with and without head are plotted. It is assumed that the first paths of the channel impulse
responses occur at the time instances, where the first significant increase of energy can be observed. The cumulative energy distributions are relatively similar for the test person and the
head phantom. However, the first paths of those channel impulse responses are delayed by only

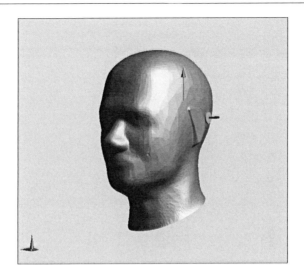

Figure 8.2: SAM head phantom used for simulations

about 600 ps compared to the reference measurement. This delay of the first paths is much smaller than the calculated delay $\Delta_t = 2.92$ ns in (8.1). Since the cumulative energy plots for the head in Fig. 8.1 show no significant increase of energy at the expected delay from (8.1) it is concluded that transmission through the head is negligible.

To verify the result that transmission through the head is negligible finite difference time domain (FDTD) simulations are performed and the attenuation for transmission through the head is calculated.

For the FDTD simulations SEMCAD [79] is used, since a model of the SAM head phantom, which was used for the measurements, is available in SEMCAD (cf. Fig. 6.2 and Fig. 8.2). The frequency range for the simulation is chosen as 1.5-8 GHz.

In Fig. 8.3, the simulated field strength in time domain is shown from the top view. The head is also plotted to see the boundaries of the field inside and outside of the head. A monopole, which is placed behind the left ear of the SAM head phantom, is used as a transmit antenna for the simulation. Position and transmit characteristic of the antenna are chosen such that the

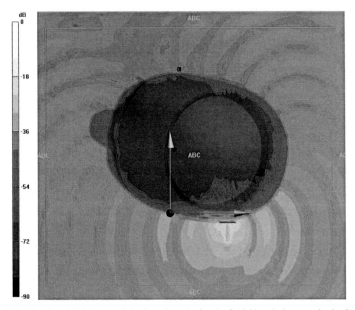

Figure 8.3: Simulated field strength in time domain for the SAM head phantom in the frequency
range between 1.5 and 8 GHz from a top view

results are comparable with the measurements.

From the simulation it can be observed that the field is strongly attenuated in direction of the
head. For the position behind the right ear, where the receive antenna was placed, an attenuation
of about 55-65 dB can be observed. This simulated attenuation corresponds with the measured
attenuation in Fig. 6.3. Although a portion of the transmitted power radiates into the head
close to the transmit antenna, it can be observed that waves radiated into the head are severely
attenuated. Thus, it is verified that direct transmission through the head is negligible.

For calculation of the attenuation through the head, the head is considered as a lossy medium.
Such a lossy medium can be described by its complex permittivity [80]

$$\begin{aligned} \varepsilon_r &= \varepsilon_r' - j\varepsilon_r'' \\ &= \varepsilon_r \left(1 - j\frac{\sigma}{\omega\varepsilon_0\varepsilon_r} \right) = \varepsilon_r \left(1 - jd_\varepsilon \right), \end{aligned} \tag{8.2}$$

with ε_r' and ε_r'' determining dispersion and losses respectively, and its complex permeability

$$\underline{\mu}_r = \mu_r' - j\mu_r'' = \mu_r \left(1 - jd_\mu\right), \tag{8.3}$$

where ε_r' and μ_r' correspond to the relative permittivity ε_r and the relative permeability μ_r of the lossless medium, respectively. Using (8.2) and (8.3) the wave number

$$\underline{k} = k' - jk'' = \omega\sqrt{\varepsilon_0\underline{\varepsilon}_r\mu_0\underline{\mu}_r} \tag{8.4}$$

becomes complex. ε_0 and μ_0 denote permittivity and permeability in free space, respectively. Thus, the electrical field strength of a wave in a lossy medium can be described as a function of the propagation distance z by

$$
\begin{aligned}
E(z,t) &= \underline{E} \cdot e^{(j(\omega t - \underline{k}z))} \\
&= \underline{E} \cdot e^{-k''z} \cdot e^{(j(\omega t - k'z))},
\end{aligned}
\tag{8.5}
$$

with the electric field-strength $\underline{E}$, the angular frequency $\omega = 2\pi f$, and the time t. In (8.5), $e^{-k''z}$ depicts the attenuation term containing

$$k'' = \frac{2\pi}{\lambda_\varepsilon} \cdot \Im\left\{\sqrt{1 - jd_\varepsilon}\right\}, \tag{8.6}$$

where $\Im\{\cdot\}$ denotes the imaginary part of the argument. The wavelength λ_ε in a lossy dielectric is given by

$$\lambda_\varepsilon = \frac{\lambda_0}{\sqrt{\varepsilon_r} \cdot \Re\left\{\sqrt{1 - jd_\varepsilon}\right\}} \tag{8.7}$$

with $\Re\{\cdot\}$ as the real part of the argument. From (8.2) one gets

$$d_\varepsilon = \frac{\sigma}{\omega\varepsilon_0\varepsilon_r}. \tag{8.8}$$

With $\lambda_0 = 0.06$ m, which corresponds to a frequency $f_0 \approx 5.2$ GHz, $\mu_0 = 4\pi 10^{-7}\frac{\text{Vs}}{\text{Am}}$, $\mu_r \approx 1$, $\varepsilon_0 \approx \frac{10^{-9}}{36\pi}\frac{\text{As}}{\text{Vm}}$ and by using the relative permittivity and conductivity for the head phantom specified in (8.6) yields

$$k'' \approx -146.4\frac{1}{\text{m}}. \tag{8.9}$$

77

Putting this result in the attenuation term $e^{-k''z}$ in (8.5) the attenuation for a distance of $z = 0.17$ m, which corresponds to the distance between the antennas on the head, becomes

$$e^{-k''z} = e^{146.4\frac{1}{\mathrm{m}}\cdot 0.17\mathrm{m}} \approx 216.2\mathrm{dB}. \tag{8.10}$$

Since relative permittivity and conductivity of the head phantom are very similar to human head tissues (cf. [81]), this result can be regarded also as valid for transmission through a human head. Calculations above show that the direct component of a transmission through the head is attenuated severely. This attenuation is much higher than expected from the transfer function plots in Fig. 6.3. Thus, it can be concluded again that energy transmitted through the head is negligible.

According to [69] direct transmission through the body is negligible and diffraction is the main propagation mechanism around the head, what we also have observed in [78]. This has strong impact on the design of antennas used in wireless BAN. Antennas should be designed in a way that almost the whole energy is radiated along the head surface to take advantage of diffraction and neither into the head nor away from the head. With such an antenna the energy that can be collected from reflections will be reduced. But this will not influence the performance of an ED or transmitted reference receiver with fixed short integration length that collects energy from paths around the head only as described above. Using antennas that do not radiate away from the head does not only reduce the energy that can be collected from reflections, but also reduces interference caused by other wireless systems in close vicinity that might prohibit UWB communication as we have shown in [82]. Since the surface of the human body looks like a plane for small areas only and it usually exhibits curvatures, dipoles or monopoles mounted perpendicular to the body surface are also suited for the use in WBANs. For such arrangements energy is radiated along possible curvatures on the body, but not directly into the body due to the antenna characteristic. To collect energy from reflections using an ED or TR receiver with adaptive integration length, antennas isotropically radiating into a half-space away from the head can be used.

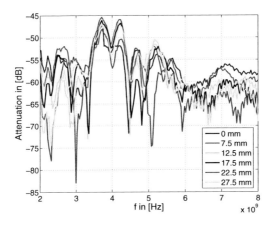

Figure 8.4: Transfer functions for different distances between antennas and head surface

8.2 Robustness Against Distance Variation of Antenna and Skin

In wireless body area networks it is hardly possible to assure a fixed distance between antenna and skin. Hence, the impact of distance variations on the channel characteristics is of great importance requiring further channel measurements moving the antennas away from the head. For several distances between the head and antennas the transfer functions were measured. In Fig. 8.4, exemplary transfer functions on the head are shown for varying distances between both antennas and the skin. From this plot it can be seen that the transfer function shapes are almost independent of the distance variation. To get a better picture of the variations with distance we have a look at the field strength attenuation for each frequency. The slopes of the field strength attenuation over the distance are fitted for each frequency by using a least-square method. These slopes of attenuation approximated from the measurements are plotted in Fig. 8.5 and exhibit severe attenuation decays only below about 3 GHz. There, the attenuation is strongly increasing with increasing distance. In the frequency range above 3 GHz, which is interesting for UWB systems, the attenuation slopes are very small, which indicates that the channel characteristics

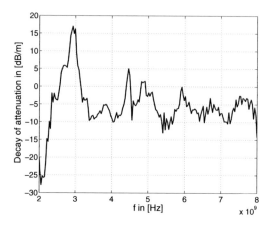

Figure 8.5: Measured decays of attenuation

is robust against small distance variations.

8.3 Reflection and Absorption

To verify the existence of any reflections or absorptions by the head, the antenna pattern for the antenna mounted on the head phantom was measured and compared with the antenna pattern without the head phantom. The antenna patterns were measured in the anechoic chamber with a VNA. The head with antenna and the antenna alone were mounted on a rotating table, respectively. The measurement setup is shown schematically in Fig. 8.6 from a top view. Please note that the antennas are not placed in the rotating axis, since the center of the head is placed at this point. The excentricity of the reference curves in Fig. 8.7 and Fig. 8.8 is caused by this head mount.

Antenna patterns measured with and without head are shown exemplarily for the frequencies 4.1 and 6.05 GHz in Fig. 8.7. For the antenna pattern measurement with the head phantom the antenna was mounted on the left ear as shown in Fig. 8.6, i.e., for $\varphi \approx 90°$ the head is located between both antennas while there exists a LOS link for $\varphi \approx 270°$. It can be observed that the

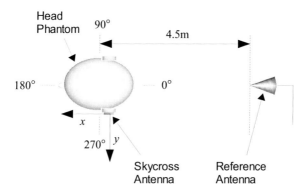

Figure 8.6: Schematic measurement setup for antenna pattern measurements with head phantom from top view

attenuation measured with the head-mounted antenna is less than for the reference measurement if an LOS link exists. This points to a reflection-like effect caused by the head. If the head acts as an obstacle between the reference antenna and the antenna mounted on the head, i.e., there exists no LOS link, the attenuation measured with the head mounted antenna is much higher compared to the reference measurement. Similar to the observations in [6], attenuations of up to 20-30 dB can be observed for such a case. Since the antenna patterns in Fig. 8.7 are only exemplary snapshots, also the energy pattern is considered to take the wideband characteristics of UWB into account. The energy pattern descriptor [83] integrates the power density radiated by the antennas over the whole time or frequency, respectively. Hence the energy pattern is defined as

$$U(\theta) = \int_{-\infty}^{\infty} |h(t,\theta)|^2 dt = \int_{-\infty}^{\infty} |H(f,\theta)|^2 df \qquad (8.11)$$

where $h(t)$ denotes the measured channel impulse response and $H(f)$ the corresponding frequency transfer function. In the following we use the frequency band between 1.5 and 8 GHz as integration limits. Since non-coherent receivers, such as energy detector or transmitted reference receiver [84], are often envisioned for the use in UWB WBANs, this energy pattern is of particular interest, because it shows the whole energy contained within the desired frequency

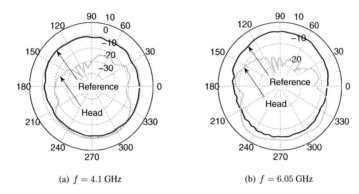

(a) $f = 4.1$ GHz

(b) $f = 6.05$ GHz

Figure 8.7: Antenna pattern with antenna directly placed to the head and reference antenna pattern without head at $f = 4.1$ GHz and $f = 6.05$ GHz

range. The normalized energy pattern in Fig. 8.8 shows a characteristic similar to the antenna patterns in Fig. 8.7. For angles between about 210 and 330° the energy pattern of the head measurement exceeds the energy pattern of the reference measurement and reflection effects can be supposed. For the remaining angles the attenuation measured with a head mounted-antenna is much higher than for the reference measurement. This means that not much energy is collected if the head is located between the reference and the head-mounted antenna.

In Fig. 8.9, the forward reflection coefficient S_{11} is shown for an antenna directly mounted on the head. For comparison, the coefficient S_{11} is also depicted for a reference measurement without head. Only slight differences between the reference and the head measurement can be observed for frequencies below 3 GHz and above 6 GHz. Between 3 and 6 GHz, the shapes of both curves are also similar, although variations of up to 10 dB can be observed. Between 2.5 and 4.6 GHz the antenna match is even better if the antenna is mounted on the head compared with the reference, i.e., more power is radiated by the antenna. From the observation in Fig. 8.7 that reflections from the head are present for LOS links and from the results in Fig. 8.9 it can be concluded that reflection and absorbtion effects exist on the human head.

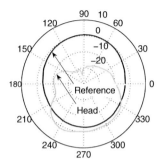

Figure 8.8: Energy pattern with antenna directly placed to the head and reference energy pattern without head

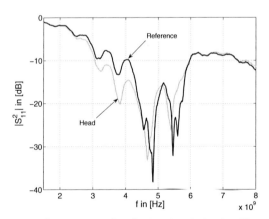

Figure 8.9: S_{11} Parameter for an antenna directly placed on the head and for an antenna without head as reference

8.4 Antenna Position

To determine the impact of the antenna position at the head several measurements with different antenna positions were performed. Both antennas were placed behind the ears, on the ears, above the ears, and in front of the ears. The measurements were done with 4 different test persons.

8.4.1 Antennas placed behind the ears

In Fig. 8.10, transfer functions are shown for both antennas placed behind the ears. The antennas were mounted perpendicular to the head with the ground plane towards the floor. In general, attenuation on the human head remains over a wide range between 50 and 60 dB. The frequency range between 3.5 and 4.5 GHz is attenuated the least, where the attenuations are between 45 and 55 dB. It can be seen that the attenuation differences for measurements on different persons are in wide ranges within 5 and 10 dB. However, there are some steep notches obvious, e.g., for person 3 at about 3.4 GHz. Since this notch only appears for one person, it does not seem to be caused by the tissue itself. Therefore, it can be assumed that there exist multiple paths around the human head which mutually interfere. Since the channel impulse responses have different amplitudes and due to the windowing in the frequency domain, we plot the cumulative energy of the CIR in Fig. 8.11 to see the energy distribution over the time. Such a plot facilitates the determination of the multipath positions in case that one tap extends to the shape of a sinc caused by the windowing in the frequency domain. It can be seen in Fig. 8.11 that the first path, which corresponds to the first strong increase of cumulative energy, is at the same position for all measurements. However, in particular the measurements on test persons 3 and 4 show a large number of dominant multipaths. These measurements show also steep notches in the transfer function plot in Fig. 8.10. CIRs for test persons 1 and 2 show only few dominant paths and the corresponding transfer functions depicts less steep notches. Moreover, it can be observed from Fig. 8.11 that almost the whole energy is contained in a window of about 2 ns duration. Since the cumulative energy plots are suited better for an analysis than the CIR plots, CIR plots are omitted in the following.

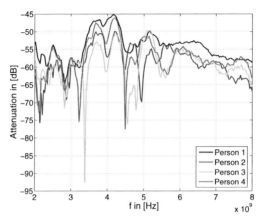

Figure 8.10: Transfer functions for antennas behind ears, down orientation, perpendicular to head

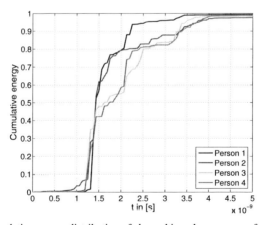

Figure 8.11: Cumulative energy distribution of channel impulse responses for antennas behind ears, down orientation, perpendicular to head

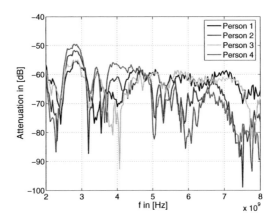

Figure 8.12: Transfer functions for antennas on ears, down orientation, perpendicular to head

8.4.2 Antennas placed on the ears

Different from the transfer function behind the ear, the attenuation lies between 60 and 80 dB if the antennas are placed directly on the ears (see Fig. 8.12). Up to 15 dB variation in attenuation are present among the different test persons. It can be seen that the transfer functions of person 1 and person 3 are very similar as well as the transfer functions of person 2 and person 4. For this position where antennas are placed directly on the ears, also some notches can be observed, probably caused by multiple paths around the head. In Fig. 8.13, the cumulative energy of the CIRs is plotted. As well as in the transfer function plot, it is also observable that the CIRs of person 1 and 3 are very similar as well as the CIRs of person 2 and 4. However, the first paths for the measurements around the head are delayed by about 0.3 ns compared to the behind ear measurements in Fig. 8.11. This is due to the increased distance between the two antennas, if both antennas are directly placed on the ears. Nevertheless, almost the whole energy is contained in a window of about 2 ns again.

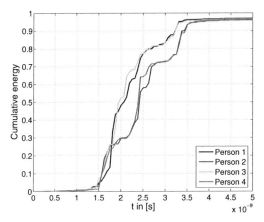

Figure 8.13: Cumulative energy distribution of channel impulse responses for antennas on ears, down orientation, perpendicular to head

8.4.3 Antennas placed in front of the ears

In Fig. 8.14, transfer functions are shown for the antenna position in front of the ear. The attenuations vary from 50 to 80 dB depending on frequency and test person. Other than the measurements behind and on the ear, the transfer functions of different persons are similar only for small frequency bands. These variations are mainly caused by two effects. The position in front of the ear is not as well defined as both previous positions. Additionally, multipath effects lead to further notches in the transfer functions. In Fig. 8.15 the distribution of the energy in the impulses responses is shown. It is obvious that not always the first path is the most dominant one. In particular, for person 4 a path delayed by about 1 ns is the most dominant one. Although the energy distributions are different for the 4 test persons, for each impulse response almost the whole energy is contained in a 2 ns window. This is the same window size as for the previously measured positions.

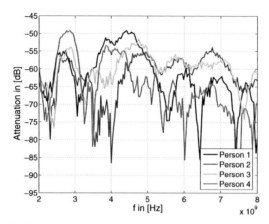

Figure 8.14: Transfer functions for antennas if front of the ears, down orientation, perpendicular to head

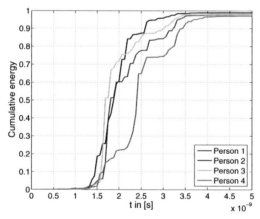

Figure 8.15: Cumulative energy distribution of channel impulse responses for antennas if front of the ears, down orientation, perpendicular to head

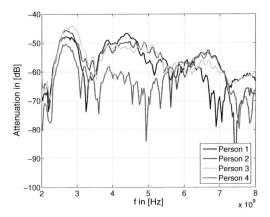

Figure 8.16: Transfer functions for antennas above ears, down orientation, perpendicular to head

8.4.4 Antennas placed above the ears

Transfer functions for the channel measurements, where the antennas were placed above the ears, are shown in Fig. 8.16. It can be seen that the transfer functions are very similar for the different test persons except for the transfer function of the test person 4, which is between 3.5 and 5.5 GHz about 10 dB worse compared to the others. These transfer functions exhibit over a wide range an attenuation of 50 to 60 dB, which is in the same order of magnitude as the attenuation if the antennas are placed behind the ears. As for the previous 3 positions also for this position some notches can be observed, which raises the assumption of multipath propagation. The position of the dominant paths can be seen in Fig. 8.17. The energy distributions are very similar for the different test persons. Again as for the previously investigated antenna positions, for this antenna position almost the whole energy is contained in a window of about 2 ns.

8.4.5 Conclusion on the different antenna positions

If all four different positions are compared together, it can be seen that attenuation is much higher if the antennas are placed in front of the ears or directly on the ears compared to the

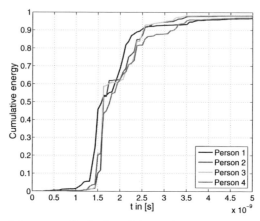

Figure 8.17: Cumulative energy distribution of channel impulse responses for antennas above ears, down orientation, perpendicular to head

above and the behind ear positions. For the above and the behind ear positions attenuations of about 50 to 60 dB were measured over a wide frequency range. However, since the behind ear position seems to be less sensitive for different persons and since this position is more suitable for an ear-to-ear communication system due to optical reasons, this position is the preferred one compared to the above ear position. Although the antenna position has a severe impact on the attenuation and the distribution of the multipaths it was observed that almost the whole energy of each channel impulse response is contained in a window of about 2 ns duration.

8.5 Integration Time in Non-coherent Receivers

According to [64] and our results in [22] the performance of transmitted reference (TR) receiver and energy detector (ED) strongly depends on the integration time used in such receivers. In the previous section it was shown that the energy is concentrated in a very short time window for the ear-to-ear link. To see the impact on an environment different from the anechoic chamber, cumulative energies of CIRs measured in anechoic chamber and office environment are plotted in Fig. 8.18. Dashed lines depict measurements, where both antennas' main directivities pointed towards the floor. For the measurements depicted by solid lines the antennas were rotated in a way that their main directivities pointed horizontally backwards. It can be observed for the anechoic chamber CIRs that almost the whole energy is contained in a window of about only 2.5 ns, i.e., the integration length is very small for an ED or TR receiver. Compared to the channel measured in the anechoic chamber, energy can also be captured from multipath reflections in the office environment. In Fig. 8.18, in particular, a path at about 6 ns is obvious.

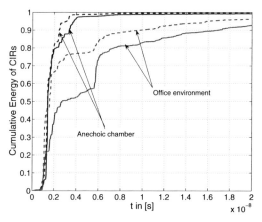

Figure 8.18: Cumulative energy of channel impulse responses in office environment and anechoic chamber; the antenna's main directivity was changed from horizontal backward (dashed lines) to vertical to the floor (solid lines)

This path is caused by a table, which the test person was sitting in front of. For the backward orientation of the antennas this path is not so strong due to the above mentioned directivity of the antennas. Nevertheless, the first paths around the head caused by diffraction can also be observed in the office environment. Using the 2.5 ns window size determined from the anechoic chamber measurements it is possible to collect about 50 and 75 % of the whole energy in the office environment, respectively, as shown in Fig. 8.18. This means that the receiver is insensitive to changes of environment by choosing a very short integration time, i.e., by only collecting energy of paths diffracting around the head.

Of course, an adaptive receiver, which adapts its integration length according to the environment, could achieve better performance. In chapter 12, a family of receiver structures with different level of channel state information (CSI) is investigated. In particular, the receiver structures with higher CSI level perform a weighting. Since very weak signal parts of the receive signal are not considered for the decision by such receivers, they can also be regarded as receivers with a variable integration time.

Part III

Interference on UWB Systems

Chapter 9

Background Noise

UWB systems are often claimed to be robust against narrowband interference [5], [13], [14] due to their large bandwidth. Since the UWB signal power is very small, this interference robustness of UWB might not always be the case. In particular for the BAN scenario, where interfering devices transmit in very close distance to the UWB receivers, interference effects could be severe. Therefore, critical interferers for UWB systems are investigated based on the standards. Additionally, measurements are carried out, since this topic is of particular importance for developing a UWB BAN system. Two types of interference are distinguished, which are in the following referred to as continuous background interference and burst interference. The term background noise is used for stationary or quasi-stationary noise. Within this definition signals from GSM basestations are also considered as background noise, since they transmit almost continuously. Frequency domain measurements of such background interferers and a solution for background interference mitigation are presented in section 9.1. The term burst interference is used for narrowband interferers (NBI) that transmit their data burst-wise such as GSM mobiles. An overview of different burst interferers as well as the results of the time-domain interference measurements are given in Section 10.1.

9.1 Background Noise Measurements

Measurements of the background noise were done in an office on the top floor of ETF building at ETH Zurich. The measurements were performed in the frequency domain using the same Skycross SMT-3TO10M antenna as for the channel measurements [66] connected to a spectrum analyzer. The antenna characteristic was calibrated out from the measurements. The resolution bandwidth of the spectrum analyzer was chosen to be 30 kHz. Since the frequency resolution for the measured frequency range is limited by the number of measurement points of the spectrum analyzer, the frequency resolution was increased by doing subsequent measurements of 20 MHz sub-bands over the whole desired frequency range between 1.5 and 6 GHz.

In Fig. 9.1, the power spectral density (PSD) averaged over measurements at different times of the day is shown. The measurements were done at ETH Zurich in the ETF building on the top floor. The next GSM and UMTS base stations are about 300 m away placed on the roof of the university hospital Zurich [85]. An overview on the interfering services, which can be observed

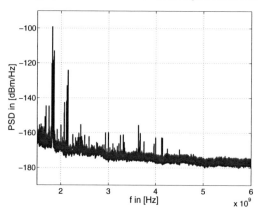

Figure 9.1: Background noise averaged over the measurements on different times of the day

in Fig. 9.1, is given in Table 9.1. Not using an additional high pass filter, an interference noise power of

$$N_{\mathrm{I}} = \int_{1.5\mathrm{GHz}}^{6\mathrm{GHz}} \mathrm{PSD}(f)df \approx -45\mathrm{dBm} \qquad (9.1)$$

Frequency range [MHz]	Service
1805 - 1880	GSM
2110 - 2170	UMTS
2400 - 2500	ISM
2700 - 3400	Radionavigation
3600 - 4200	Fixed wireless services

Table 9.1: Interfering services observed from background noise measurements in Fig. 9.1

is present in the frequency range between 1.5 and 6 GHz. This background noise power is dominated by the interference from GSM and UMTS base stations, which can be seen in Fig. 9.1.

In Fig. 9.2, PSDs of the background noise measurements at different times of the day are shown. It can be observed that GSM and UMTS base stations are the main interferers at any time. Only the measured power spectral density of GSM changes for the different times of the day. This depends on the load of the base stations during the different measurements. For UMTS, which does not use burst wise transmission in frequency division duplex (FDD) mode, the measured power spectral densities are almost the same for measurements carried out at different times of the day. Besides these both interferers, some less dominant interferers could have been measured in the frequency range between about 3 and 4.5 GHz. Since these interferers were not present in all measurements, it can be assumed that these interferers have varying loads at different times of the day.

97

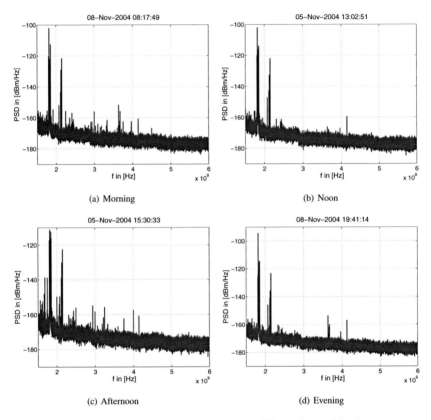

(a) Morning

(b) Noon

(c) Afternoon

(d) Evening

Figure 9.2: Background noise measured at different times of the day

Chapter 10

Burst Noise

In the previous section it has been shown that GSM and UMTS base stations cause the dominant background noise. However, there exists a number of other interferers. In particular, interferers that are located in close vicinity to the UWB receiver might be harmful. If the instantaneous signal power of an interferer is too high, it is assumed that the UWB receiver suffers from clipping and no UWB transmission is possible.

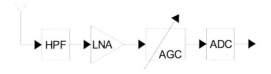

Figure 10.1: Receiver model for clipping considerations

For the clipping considerations the receiver model in Fig. 10.1 is assumed, consisting of a high pass filter (HPF), a low noise amplifier (LNA), an automatic gain control (AGC), and an analog-to-digital converter (ADC) with 6 bit resolution. It is assumed that 3 bit will be used for the desired UWB signal, while the remaining 3 bit are a reserve for noise and interference until the receiver suffers from clipping. Since the noise $w(t)$ is added by the LNA, the HPF does not limit the noise bandwidth, but only attenuates background interference. The LNA and the AGC are assumed to be perfect and they do not cause any clipping. The AGC amplifies the desired UWB receive signal in such a way that it fits best in the desired 3 bit range of the ADC. However, if the input signal of the ADC exceeds the 6 bit due to an interference signal, the ADC suffers from clipping and no UWB signal can be resolved. Otherwise, it is assumed that the UWB signal can be detected by signal processing if the ADC is not clipping. Based on the assumptions above, the ADC is clipping for any noise or interference signal that is larger than 7 the desired signal amplitude.

In Fig. 10.2, signal amplitudes of Bluetooth (BT), GSM 1800, and IEEE 802.11b WLAN are shown for distances up to $d = 5$ meters. Each system transmits with its maximum allowed transmit power. Only free space attenuation is assumed for the interferers and a 5th order Butterworth high pass with $f_l = 3.1$ GHz in the UWB receiver. For bandwidths up to about 10% of the center frequency, resonant circuit designs can be used in the analog part of the receiver. Since such resonant circuits allow for energy efficient hardware realizations, a system having the minimum UWB bandwidth of 500 MHz is assumed in the remainder. Considering the transmit power spectral density of $\text{PSD}_{\text{TX},50} = 0\text{dBm}/50\text{MHz}$, which is maximally allowed by the FCC, and an attenuation of 60 dB from ear to ear, the receive signal power becomes $P_{\text{RX},500} = -50$ dBm, which corresponds to a signal amplitude of 0.7 mV @ 50 Ω. Using the assumption that the ADC is clipping if the interferer signal is 7 times higher than the desired signal, the clipping limit is 4.9 mV. From Fig. 10.2 it can be seen that this clipping limit is exceeded by the interferers amplitudes only for distances below 1 m. Thus, only devices in the direct environment of a person are possible interferers.

Considering different burst interferers, interferers with periodic burst structures and non-periodic burst structures are distinguished. In Fig. 10.3, normalized time domain signals of GSM, BT, and IEEE 802.11b WLAN are depicted. As expected from the standards, BT and GSM show a periodic burst structure while WLAN does not exhibit such a periodicity. Nevertheless, for all burst interferers segments between adjacent bursts can be observed, where the channel is idle.

In the following, it is assumed that the channel can have the two different states: occupied or idle. If the channel is occupied, the ADC is assumed to be clipping and no UWB transmission is possible, i.e., UWB transmission is only possible if the channel is idle. This is shown

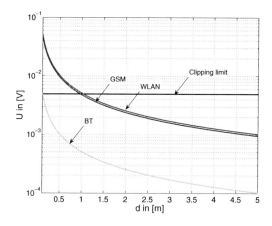

Figure 10.2: Signal amplitudes of interferers considering free space attenuation

exemplarily in Fig. 10.4. To determine the idle times we measure the interference with a real time sampling oscilloscope. For these measurements, interferer and measurement antenna were placed close to each other. Since the interfering signal shall not be reconstructed and since we are interested only in the duration and position of the interfering bursts, the sampling frequency was chosen as 1 MSample/s.

10.1 Measurement Results of Burst Interferers with Fixed Burst Lengths

In this section we focus on interferers that transmit their data periodically in bursts of fixed length. This group of interferers consists of the global system for mobile communications (GSM), the digital enhanced cordless telecommunications (DECT) system, Bluetooth (BT), and microwave ovens. While dedicated frequencies can be used by GSM and DECT, Bluetooth and microwave ovens are using the non-licensed industrial, scientific and medical (ISM) band around 2.4 GHz.

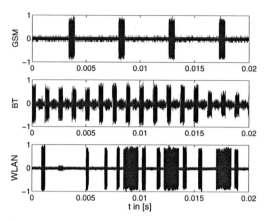

Figure 10.3: Measured time domain signals of GSM, BT, and WLAN with normalized amplitudes

Figure 10.4: 2-state interference model; state 1 if an interferer burst is present and state 0 if no interferer burst is present

10.1.1 GSM

The GSM system uses frequency bands at 900 and 1800 MHz [86]. To distinguish both frequency bands, GSM systems operating in the respective band are also referred to as GSM900 and GSM1800. From Tab. 10.1 it can be seen that up- and downlink are separated by 45 MHz in GSM900 and by 95 MHz in GSM1800. In GSM900 there are 174 carriers and in GSM1800 374 carriers used, each spaced by 200 kHz. Maximum transmit powers of the mobile devices are 2 W (33 dBm) for GSM900 and 1 W (30 dBm) for GSM1800.

As mentioned above, bursts are transmitted in GSM periodically in a time division multiple access (TDMA) scheme [87]. One TDMA frame consists of eight time slots, each of them having duration of $\frac{3}{5200} \approx 577$ μs. Thus, the duration of a GSM burst is 577 μs. The repetition period of a time slot is 4.62 ms, which corresponds to the duration of a TDMA frame. In

Fig. 10.5, a GSM burst structure measured in time domain is shown exemplarily. The measured burst lengths correspond with the ones expected from the standard. Using the general packet radio service (GPRS) for data transmission the general burst structure remains as described above. However, mobile and base stations can transmit with GPRS in up to 8 adjacent bursts to increase the data rate [88], which reduces the time during which the channel is not occupied.

	GSM900	GSM1800
Mobile	880-915 MHz	925-960 MHz
Base station	1710-1785 MHz	1805-1880 MHz

Table 10.1: GSM frequency ranges

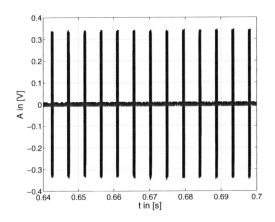

Figure 10.5: Measurement of GSM bursts

According to the evaluation procedure described in the previous section (cf. Fig. 10.4) we do not consider now the amplitudes but only the two states "interferer present" and "no interferer present". Based on these two states we determine the measured burst durations. In Fig. 10.6(a), the histogram for GSM burst durations is plotted. From the standard there would be one peak at 576 μs expected. The measured burst lengths are slightly below this value due to the tuning at the beginning and at the end of each burst. The histogram for the time between two GSM burst, which we refer to as idle time, can be seen in Fig. 10.6(b). The strong peak at about 4ms corresponds well with the expected 7 times 576 μs caused by the 7 empty GSM slots. The second peak at about 8.6 ms is also expected from the standard. 8 GSM slots form one TDMA-frame and 26 TDMA-frames one multi-frame. One of these 26 TDMA-frames is an idle frame, where no transmission takes place. This idle frame allows the mobile device some space for housekeeping, e.g, signal strength measurements.

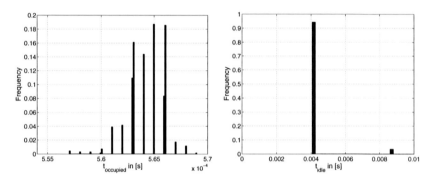

(a) Histogram of the measured burst lengths of one GSM interferer

(b) Histogram of the measured channel idle times in presence of one GSM interferer

Figure 10.6: Histograms of GSM burst structures and channel idle times

10.1.2 DECT

The whole frequency range defined for DECT reaches from 1880 MHz to 1980 MHz and from 2010 MHz to 2025 MHz [89]. Additionally, the 902 MHz to 928 MHz ISM band and the 2400 MHz to 2483.5 MHz ISM band are defined for DECT use in the USA. In Europe the spectrum between 1880 and 1900 MHz is used for DECT. In this frequency band there exist 10 different carriers. The center frequencies of these carriers are given by

$$f_c = f_0 - c \cdot 1.728 \, \text{MHz} \tag{10.1}$$

with $f_0 = 1897.344$ MHz and $c = 0, 1, \ldots, 9$. The maximum transmit power of a DECT device is 250 mW (24 dBm).

The medium access is done by a TDMA-scheme with 24 full time-slots [90]. The first 12 time-slots are used for the downlink, while the second 12 time-slots are used for the uplink. These 24 time-slots form one frame and 16 frames form one multiframe. One frame has the length 10 ms. Therefore, one full time-slot has a duration of about 417 μs. In Fig. 10.7, a measured DECT burst structure can be seen in time domain. The measurement corresponds well with the burst durations expected from the standard.

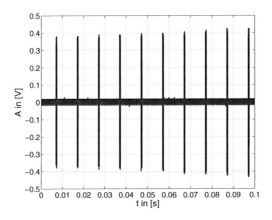

Figure 10.7: Measurement of DECT bursts

In regards to GSM, burst lengths depicted in Fig. 10.8(a) are also slightly shorter than expected from the standards. This effect is again caused by the tuning of the burst, which is in the same order of magnitude as the noise.

The measured idle time between two adjacent DECT bursts, which is shown in Fig. 10.8(b), corresponds well with the values from the standard. The measured idle times are about 9.61 ms while the expected are about $23 \cdot 417 \mu s = 9.59$ ms.

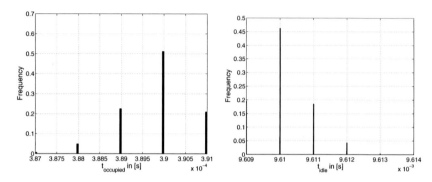

(a) Histogram of the measured burst lengths of one DECT interferer

(b) Histogram of the measured channel idle times in presence of one DECT interferer

Figure 10.8: Histograms of DECT burst structures and channel idle times

10.1.3 Bluetooth

Unlike GSM and DECT, no dedicated frequency band is used for Bluetooth [91]. For Bluetooth, the 2.4 GHz ISM band is used for transmission. The frequency band starts at 2400 MHz and ends at 2483.5 MHz. There are 79 different channels with 1 MHz spacing defined according to

$$f = 2402 + k \text{ MHz}, k = 0, \dots, 78. \tag{10.2}$$

For Bluetooth, 3 different power classes are defined. The maximum output power is 100 mW (20 dBm) for devices of class 1, 2.5 mW (4 dBm) for devices of class 2, and 1 mW (0 dBm) for devices of class 3.

The burst repetition period in Bluetooth is $1250~\mu$s. This time interval is formed by one transmit and one receive slot, each lasting for $625~\mu$s. According to [91], there exist single-slot and multi-slot packets. Multi-slot packets can occupy either 3 or 5 slots, i.e, they can have a duration up to 1875 and 3125 μs, respectively. In single-slot packets, only the first 366 μs of a slot are occupied by a Bluetooth burst and the remaining 259 μs are required for tuning the frequency synthesizer to the adjacent frequency band.

A Bluetooth network is formed by one master and up to seven slaves in active mode. Hence, if a Bluetooth network consists only of the master and one slave, both devices are transmitting alternately. If there exists more than one slave, only the master is transmitting in every second slot. The slaves transmit alternately depending on which slave is polled by the master node.

For the time domain measurement, a Bluetooth network with master and one slave is considered. In Fig. 10.9, the measured burst structure is shown. It can be observed that only single-slot packets are present. However, the Bluetooth signal slightly differs from the burst structure described in the standard. Although burst repetition rate and burst duration correspond, only a burst in the transmit slot can be observed in the measurement. In the receive slot, no burst is measured since this device was placed further away from the measurement antenna. Due to the low transmit power of a Bluetooth headset, which was used for this measurement, only the signal of one device was detected.

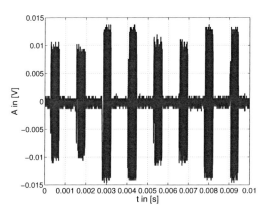

Figure 10.9: Measurement of BT bursts

The histogram of the BT burst length is given in Fig. 10.10(a). Different from the previously shown GSM and DECT histograms there exist more than one peak, i.e., only one burst length. For this measurement, we used a BT headset, which was placed close to the UWB antenna, and a BT mobile, which was placed further away. Fig. 10.9 shows a part of the measured signal where only the headset bursts are obvious. But there exists also parts, where the signal from the mobile is larger than the noise. Such signal parts can be above the threshold for the decision if the channel is occupied or not, and lead to the burst lengths different from the standards, especially to the very short ones.

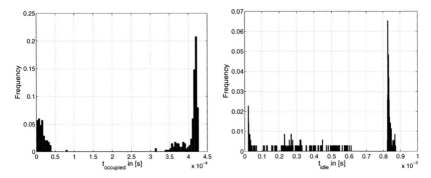

(a) Histogram of the measured burst lengths of one BT interferer

(b) Histogram of the measured channel idle times in presence of one BT interferer

Figure 10.10: Histograms of BT burst structures and channel idle times

A similar observation can be made for the idle times depicted in Fig. 10.10(b). Many different idle times are detected but there exists one peak at about 850 μs. This peak corresponds with the expected idle time of $2 \cdot 625\ \mu\text{s} - 366\ \mu\text{s} = 884\ \mu\text{s}$ between two adjacent BT bursts.

10.1.4 Microwave Ovens

For microwave ovens there exist no dedicated standards. These devices only have to fulfill the regulations for 2.4 GHz ISM devices. In Fig. 10.11, a time domain measurement of a signal generated by a microwave oven is shown. It can be observed that a burst has a duration of 10 ms. The burst repetition period is 20 ms. It has to be noted that these values cannot be regarded as general. For different types and realizations of microwave ovens there exist different burst structures as shown in [92].

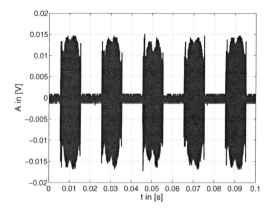

Figure 10.11: Measurement of microwave oven bursts

10.2 Measurement Results of Burst Interferers with Variable Burst Length

10.2.1 IEEE 802.11b

IEEE 802.11b is a widely used standard for wireless local area networks (WLAN) [93]. As many other wireless systems, IEEE 802.11b devices use the 2.4 GHz ISM band for transmission. In Europe, 13 different channels are defined with center frequencies

$$f = 2412 + k \cdot 5 \, \text{MHz}, k = 0, \dots, 12. \tag{10.3}$$

Each channel has a bandwidth of $B = 22$ MHz. Therefore, in the 2.4 GHz ISM band there is a maximum of 3 orthogonal frequency bands possible, i.e., a maximum of 3 IEEE 802.11b networks can be operated in parallel without mutually interfering. The maximum allowed transmit power in Europe is 100 mW (20 dBm) and in the USA 1000 mW (30 dBm).

Different from the wireless systems described in the previous section IEEE 802.11b bursts have variable length and do not exhibit a periodic structure. The MAC for all IEEE 802.11 derivatives and thus also for IEEE 802.11b is defined in [94]. Since there exist different realizations and data rates for different IEEE 802.11 systems, burst lengths are defined in octets (= 8 bit). Therefore, the burst lengths depend on the data rate as well as on the payload. The payload length can vary between 0 and 2312 octets. An exemplary time domain measurement, which exhibits this variable IEEE 802.11b burst structure, is shown in Fig. 10.12.

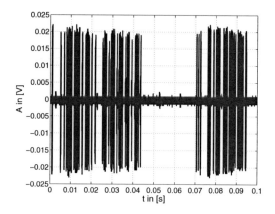

Figure 10.12: Measurement of WLAN bursts

A histogram of the WLAN burst lengths is given in Fig. 10.13(a). It can be observed that there exist different burst lengths with a duration of up to about 1.4 ms. But most WLAN bursts have durations of about 1.3, 0.3, 0.2, or less than 0.1 ms. In most cases the channel is not occupied for less than 2ms, which can be seen in Fig. 10.13(b). Nevertheless, the channel might be also idle for a time period of up to 10 ms.

10.2.2 UMTS

For the Universal Mobile Telecommunications System (UMTS) two different transmission modes are distinguished [95] to allow for efficient spectrum usage according to the frequency regulations in different regions. In the frequency division duplex (FDD) mode two different frequencies are used for the up- and downlink. In the time division duplex (TDD) mode the same frequency but different time slots are assigned for the up- and downlink. This possibility to use

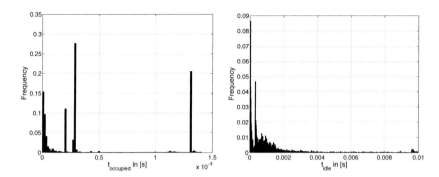

(a) Histogram of the measured burst lengths of one WLAN interferer

(b) Histogram of the measured channel idle times in presence of one WLAN interferer

Figure 10.13: Histograms of WLAN burst structures and channel idle times

either FDD or TDD mode allows for an efficient spectrum usage according to the frequency allocation in different regions [95].

FDD For the FDD mode six different operating bands are defined [96]. The operating bands with corresponding frequency bands are given in Tab. 10.2. In Switzerland, operating band

Operating band	Uplink frequencies [MHz]	Downlink frequencies [MHz]
I	1920-1980	2110-2170
II	1850-1910	1930-1990
III	1710-1785	1805-1880
IV	1710-1755	2110-2155
V	824-849	869-894
VI	830-840	875-885

Table 10.2: Operating bands for UMTS in FDD mode [96]

I is used for UMTS nowadays [97]. The UMTS channel spacing is 5 MHz and the channel bandwidth amounts to 5 MHz, too [95], which results in 12 channels for operating band I.

The maximum UMTS transmit power of the mobile, also referred to as user equipment (UE), depends on the power class of the device. Except from operating band 1 the maximum transmit powers are either 250 mW (24 dBm) or 125 mW (21 dBm). Additionally, maximum transmit powers in band 1 can also be 2 W (33 dBm) and 500 mW (27 dBm).

In FDD mode one frame consists of 15 slots. Since one frame has a duration of 10 ms each slot has a length of about 667 μs. In general, UMTS FDD transmission is continuous from the instant when it has started to the moment when it has stopped [98]. Different users are separated by different codes using wideband code division multiple access (WCDMA). However, transmission gaps are possible in compressed mode [99]. These transmission gaps are used for the measurements on different carrier frequencies during inter-frequency handovers of the UE. Up to 7 slots, i.e., about 4.67 ms, can be used for the transmission gap. There are three different methods to achieve the compressed mode, i.e, to reduce the transmission time. (i) A rate matching is achieved by puncturing, (ii) the spreading factor is decreased by a factor of 2 to increase the data rate, or (iii) the number of transmit bits is reduced by higher layer scheduling. Since our time domain measurements were performed in a static environment, no handovers were necessary and no gaps were observed from the time domain measurements (see Fig. 10.14).

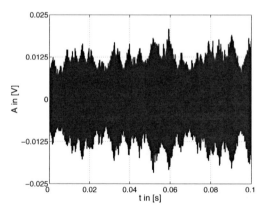

Figure 10.14: Measurement of UMTS bursts

TDD   In most countries the FDD mode is used for UMTS transmission. Nationwide coverage by TDD UMTS networks exists only in the Czech republic, Lithuania, Malaysia, and New Zealand [100]. However, TDD UMTS is used in several other countries in some cities or regions, partially for trial purposes. The operating bands for UMTS in TDD mode are given in Table 10.3. The operating bands b) and c) are used in ITU region 2, i.e, in North and South America, and additional frequency bands in region 2 are for further studies. However, it is not precluded to use existing or other frequency bands for TDD UMTS. In TDD UMTS, two different options, which yield different chip rates, are distinguished; the 3.84 Mcps option and the 1.28 Mcps option. The channel spacings are 5 MHz for the 3.84 Mcps option and 1.6 MHz for the 1.28 Mcps option [101]. The maximum output power for the power classes 1 to 4 are 1 W (30 dBm), 250 mW (24 dBm), 125 mW (21 dBm), and 10 mW (10 dBm), respectively.

Operating band	Uplink frequencies [MHz]	Downlink frequencies [MHz]
a)	1900-1920	2010-2025
b)	1850-1910	1930-1990
c)	1910-1930	1910-1930

Table 10.3: Operating bands for UMTS in TDD mode [101]

According to [102] both options for TDD mode have different burst structures.

Using the 3.84 Mcps option, a frame of 10 ms duration contains 15 slots as in the FDD mode. Hence, each slot has a duration of 667 μs. The slots are used either for the uplink or for the downlink. There is no special allocation of the slots for up- or downlink in [102]. The only restriction is that at least one time slot has to be assigned for the downlink and one for the uplink. Therefore, using the 3.84 Mcps option, the UE does not transmit for at least 667 μs up to at most 9.33 ms during a frame of 10 ms duration.

The 10 ms frame is divided into 2 subframes of 5 ms duration in the 1.28 Mcps option. Each subframe contains 7 time slots of 675 μs duration and additional 352 chips that are reserved for downlink and uplink pilot time slots and the main guard period. Different from the 3.84 Mcps option in the 1.28 Mcps option the first time slot is always allocated as downlink while the second time slot is always allocated as uplink. Thus, in a frame of 10 ms duration the UE does not transmit for at least $2 \cdot 675 \ \mu s = 1.35$ ms up to at most 8.65 ms.

These burst structures have not been verified by measurements, since, as mentioned above, the UMTS TDD mode is only used in very few countries.

Chapter 11

Impact of the Interference Results on the System Design

11.1 Impact of Background Interference

11.1.1 Link Budget for Thermal Noise

For the purpose of a short recapitulation the link budget, which assumes thermal noise only, is calculated. UWB transmission in the FCC peak power limit is assumed. The peak power limit is defined by a maximum transmit power spectral density of

$$\text{PSD}_{\text{TX},50} = 0 \text{ dBm}/50 \text{ MHz}, \tag{11.1}$$

allowing a maximum pulse repetition rate of 1 MHz [103]. According to the same publication the energy density of a single pulse E_{sp} is given by

$$\begin{aligned} E_{sp} &= \frac{0.45^2 \cdot P_p}{2 B_\mu^2} \\ &= 4.05 \cdot 10^{-20} \frac{\text{W}}{\text{Hz}^2}, \end{aligned} \tag{11.2}$$

with the peak power $P_p = 1 \text{ mW}$ and the measurement bandwidth $B_p = 50 \text{ MHz}$. This yields for a pulse with a bandwidth of $B = 500 \text{ MHz}$ a peak energy E_p of

$$E_p = 2.025 \cdot 10^{-11} \text{J} \equiv -77 \frac{\text{dBm}}{\text{Hz}}. \tag{11.3}$$

Considering the ear-to-ear link the attenuation is assumed to be PL $= 60$ dB. Assuming no further losses due to antenna mismatch or imperfect components the receive energy is then given by

$$E_{\text{RX}} = -137 \text{ dBm/ Hz.} \tag{11.4}$$

First, we consider thermal noise as the only noise source. For a temperature of 293 K the noise power spectral density becomes

$$N_0 = -174 \frac{\text{dBm}}{\text{Hz}}. \tag{11.5}$$

Hence, the ratio of receive signal energy E_{RX} and noise power spectral density N_0 is given by

$$\frac{E_{\text{RX}}}{N_0} \approx -137 \frac{\text{dBm}}{\text{Hz}} + 174 \frac{\text{dBm}}{\text{Hz}} = 37 \text{ dB.} \tag{11.6}$$

Please note that no averaging of UWB pulses is considered. From (11.6) it can be seen that this $\frac{E_{\text{RX}}}{N_0}$ is high enough for communication assuming thermal noise only. Since the $\frac{E_{\text{RX}}}{N_0}$ is not only determined by thermal noise but also by interference, a link budget based on background noise measurements presented in Section 9.1 is calculated as well.

Link Budget for Background Noise

Considering the background noise from (9.1) the $\frac{E_{\text{RX}}}{N_I}$ reduces to

$$\frac{E_{\text{RX}}}{N_I} \approx -137 \frac{\text{dBm}}{\text{Hz}} + 132 \frac{\text{dBm}}{\text{Hz}} = -5 \text{ dB,} \tag{11.7}$$

where N_I accounts for the thermal noise as well as for the background noise. The ratio $\frac{E_{\text{RX}}}{N_I}$ is about 30 dB below $\frac{E_{\text{RX}}}{N_0}$, where only thermal noise is considered. Since antennas have an implicit band pass characteristic, it would be desirable to use the antenna as a band pass only. By this way, an additional filter could be avoided. If we consider only the Skycross antenna, which was used for the measurements, as a filter, the interference noise power spectral density reduces to

$$N_I \approx -142 \frac{\text{dBm}}{\text{Hz}}, \tag{11.8}$$

and a ratio $\frac{E_{RX}}{N_I}$ of

$$\frac{E_{RX}}{N_I} \approx -137 \, \frac{\text{dBm}}{\text{Hz}} + 142 \, \frac{\text{dBm}}{\text{Hz}} = 5 \, \text{dB} \qquad (11.9)$$

can be achieved. From (11.7) it can be seen that the achieved $\frac{E_{RX}}{N_I}$ is not large enough for reliable impulse radio communication and that especially the interference from GSM and UMTS basestations has to be mitigated. This can be done either by applying an antenna, whose transfer function has steep slopes, or an additional filter that attenuates out-of-band interferers.

To determine the impact of such a filtering on the interference, Butterworth high pass filters of different orders are considered. Since a UWB device for BAN applications should be as simple as possible and since the path loss increases with increasing frequency, the desired frequency band should be located on the lower end of the UWB spectrum. Considering the FCC allowed UWB frequency range $3.1 - 10.6$ GHz we choose the lower 3 dB cut-off frequency of the Butterworth filter as $f_l = 3.1$ GHz. In Table 11.1 background noise power spectral densities based on the interference measurements from $1.5 - 6$ GHz are shown for different filter orders. With increasing filter order a reduction of the noise power spectral density and a simultaneous $\frac{E_{RX}}{N_I}$ increase can be observed. Using a 5th order Butterworth high pass, the noise power spectral density is about -167.8 dBm/Hz. The noise power spectral density due to thermal noise only is -174 dBm/Hz. Thus, with the 5th order Butterworth high pass the noise power is about 6 dB higher than the noise power for thermal noise, only. Note that the desired signal is also attenuated by the filter in the pass band range. However, this attenuation does not exceed 3 dB.

Filter order	1	2	3	4	5
Noise power spectral density [dBm/Hz]	-150.7	-154.9	-159.4	-163.8	-167.8

Table 11.1: Background noise power for Butterworth high pass filters of different orders

11.2 Impact of Burst Interference

In the Section 10.1, the interferer burst structures given in the standards were verified by means of time domain measurements. It could have been observed that GSM, DECT, Bluetooth, and the microwave oven have fixed burst durations, while the burst durations of IEEE 802.11b are variable. Different from the above mentioned systems that use TDMA, UMTS in FDD mode uses CDMA. Therefore, UMTS devices can be transmitting continuously yielding bursts with duration of the active connection. Since all interferers have a transmit power much higher than UWB, which can be seen in Tab. 11.2, no UWB transmission is possible if such interferers are in very close distance to the UWB device assuming the clipping model, which was presented at the beginning of this section. To avoid such interference we propose a UWB transmission between adjacent interferer bursts, which we refer to as temporal cognitive UWB medium access. A detailed description is given in Chapter 13. Unfortunately, the temporal cognitive UWB medium access is not suited to avoid UMTS FDD interference. This interference has to be avoided by a receive bandpass filter with a steep edge or by a notch filter that attenuates signals within the UMTS frequency bands.

Standard	Frequency range [GHz]	Number carriers	Channel spacing [MHz]	Bandwidth [MHz]	Burst length [μs]	Burst repetition period [μs]	Transmit power [mW]
GSM900	0.88-0.915 0.925-0.96	174	0.2	0.2	577	4615	2000
GSM1800	1.71-1.785 1.805-1.88	374	0.2	0.2	577	4615	1000
DECT	1.88-1.9	10	1.728	1.728	417	10000	250
Bluetooth	2.4-2.4835	79	1	1	366	625	1/ 2.5/ 100
IEEE 802.11b	2.4-2.4835	13	5	22	var.	var.	100
UMTS	1.92-1.98 2.11-2.17	12	5	5	-	-	125/ 250
Microwave oven	2.4	-	-	-	10000	20000	-

Table 11.2: Overview of the wireless systems considered in chapter 11

Part IV

Physical Layer

Chapter 12

Receiver Structures

Different from narrowband systems, many multipaths from the environment can be observed in UWB systems due to the wider bandwidth. This makes a channel estimation, which is necessary for coherent receiver structures, very complex. Hence, coherent receiver structures, such as the RAKE receiver that was presented in Section 4.7, are not suited for the use in WBANs. Furthermore, the non-coherent receiver structures seem to be more promising for applications requiring low complexity. Two non-coherent receiver structures having a very similar structure are the energy detector and the transmitted reference receiver, which are both shown in Section 4.7. Since there is a number of publications on transmitted reference receivers and energy detectors, only an overview of the most relevant publications is given in the following. Both receiver structures were introduced in the 1960s for the first time [15], [16] but started again to raise interest with the beginning of UWB communications. In [17], the transmitted-reference receiver was presented for the first time in conjunction with UWB communications including a delay hopping, i.e., different delays are used for different users. Moreover, an expression for the error probability was derived. The multi-user capabilities of this delay-hopped transmitted-reference system were shown in [104]. Since the reference pulse only depicts a noisy template for correlation in the receiver, a maximum-likelihood estimator for the template signal was presented in [105] yielding a substantial performance improvement. An improvement of the template signal is done in [18] by averaging over several noisy reference signals. Moreover, a generalized likelihood ratio test for transmitted reference systems is presented as well as a differential transmitted reference system, where the previous data signal serves a reference signal. A similar averaging over the reference signals is proposed in [19] to reduce the impact of narrowband interference on the receiver. In [20], such an averaging of the reference signals is shown to be optimal considering no channel knowledge and not restricting on one doublet only for decision.

To overcome the drawbacks of transmitting one reference pulse for each data pulse in [106] a pilot aided system was proposed, where only every k data pulses one reference pulse is transmitted. There, the classical transmitted reference system is regarded as one special case of the pilot aided system. In [21], a scheme is presented, where two bits are transmitted using one doublet. One bit determines the position of the reference pulse while the second bit determines the amplitude of the data bit. Conceptually, this is the same scheme as our transmitted-reference pulse interval amplitude modulation (PIAM), where the data is modulated not only in the amplitude but also in the delay [22]. Hence, the performance of both schemes is equal. Just by adding a pulse amplitude modulation (PAM) for the reference pulse due to spectrum shaping reasons, the same scheme was presented in [107]. Another way to improve the performance of transmitted-reference systems was presented almost in parallel in [23] and in [24]. There, a weighting after multiplication of the reference and the data signal is proposed. The calculation of the bit error probability for such a weighting scheme was shown in [25] considering inter-symbol interference. A closed form expression for the error probability of a transmitted-reference system was derived in [26]. Additionally, a lower bound for the bit error probability was calculated in [27]. In [28], an average log-likelihood ratio test for transmitted reference systems was derived assuming Rayleigh and lognormal path strength models. Based on these optimum receiver structures also some suboptimal receivers were obtained.

In Section 12.2 and 12.3, maximum likelihood receivers assuming different levels of channel state information are derived for binary PPM and transmitted reference PAM in the presence of inter-symbol interference. OOK is not considered in the following due to the required decision threshold, whose determination increases the complexity compared to a receiver for binary PPM. However, there are ways to determine the threshold such as the iterative approach presented in [108]. Besides the derivation of the above mentioned transmitted reference receivers, the maximum likelihood transmitted reference receivers are derived for the case of a present co-channel interference. There, also partial channel state information is assumed. These calculations are shown in Section 12.4. The performance of the achieved receiver structures is compared by means of simulation. In Section 12.5, maximum likelihood receivers for the above mentioned transmitted reference pulse interval amplitude modulation concept are derived for different CSI level. Finally in section 12.6, the performance of the derived receiver structures for binary PPM, transmitted reference PAM, and transmitted reference PIAM are compared for the interference free case. The results shown in this section were partially presented in [22], [109], and [110].

12.1 Maximum-A-Posteriori Rule

The derivation of the maximum likelihood receiver structures is based on the maximum-a-posteriori (MAP) rule [54]. Using the MAP rule, the probability that a signal $\vec{s}$ has been transmitted is maximized under the condition that a signal $\vec{r'}$ is received assuming a certain CSI, i.e.,

$$\hat{\vec{s}} = \arg \max_{\vec{s}_i} P[\vec{s} = \vec{s}_i | \vec{r'}, \text{CSI}]. \tag{12.1}$$

The term to be maximized in (12.1) can be rewritten as

$$P[\vec{s} = \vec{s}_i | \vec{r'}, \text{CSI}] = \frac{p(\vec{s} = \vec{s}_i, \vec{r'}, \text{CSI})}{p(\vec{r'}, \text{CSI})}. \tag{12.2}$$

Applying the Bayes rule [111], one obtains

$$P[a = a_i | \vec{r'}, \text{CSI}] = \frac{p(\vec{r} | \vec{s} = \vec{s}_i, \text{CSI}) \cdot P[\vec{s} = \vec{s}_i, \text{CSI}]}{p(\vec{r'} | \text{CSI}) \cdot P[\text{CSI}]}. \tag{12.3}$$

In the cases considered in the following, the transmitted signal and the channel state information are statistically independent. Thus,

$$P[\vec{s} = \vec{s}_i, \text{CSI}] = P[\vec{s} = \vec{s}_i] \cdot P[\text{CSI}] \tag{12.4}$$

and (12.3) becomes

$$P[\vec{s} = \vec{s}_i | \vec{r'}, \text{CSI}] = \frac{p(\vec{r'} | \vec{s} = \vec{s}_i, \text{CSI}) \cdot P[\vec{s} = \vec{s}_i]}{p(\vec{r'} | \text{CSI})}. \tag{12.5}$$

Since the denominator in (12.5) is independent of $\vec{s}_i$, only the numerator of (12.5) is considered for the derivation of the ML estimators, i.e.,

$$P\left[\vec{s} = \vec{s}_i \middle| \vec{r}', \text{CSI}\right] \propto p\left(\vec{r}' \middle| \vec{s} = \vec{s}_i, \text{CSI}\right). \tag{12.6}$$

According to (12.6) the whole transmit signal $\vec{s}$ is estimated based on the whole receive signal $\vec{r}$. Such an ML receiver would require a buffer at the receiver side as well as the processing of the whole data, which both increase the complexity of the receiver structures. Since the receiver structures for BAN applications shall be of low complexity, detection is done symbolwise for each symbol a. As shown in Fig. 12.1, the decision is based on the receive signal $\vec{r} = [\vec{r}_1, \vec{r}_2]$ within one symbol interval of duration T. Thus, in the remainder

$$P\left[a = a_i \middle| \vec{r}, \text{CSI}\right] \propto p\left(\vec{r} \middle| a = a_i, \text{CSI}\right). \tag{12.7}$$

is used for the derivation symbolwise ML detectors.

12.2 Maximum Likelihood Receivers for Binary PPM in the Presence of ISI

12.2.1 System Model

It is assumed that only one pulse is transmitted per symbol, which is a reasonable assumption for short range communication in wireless body area networks due to the moderate path loss. Although the symbol-wise ML detector bases its decision only on one single PPM frame, for the derivation, the previous receive signal has also to be taken into account as it may cause ISI as shown in Fig. 12.1. We assume in the remainder of this section that the maximum duration of the channel impulse response is smaller than the duration of a PPM frame T, i.e., the ISI has only an impact on the next PPM half-frame. Time-hopping is omitted due to clarity of the derivation but can be easily included. Thus, the sampled receive signal in the considered PPM frame is given by

$$\vec{r}_1 = \frac{1}{2}(1 - a_1)\vec{h} + \frac{1}{2}(1 + a_0)\vec{g} + \vec{n}_1 \tag{12.8}$$

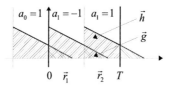

Figure 12.1: Schematic of the considered receive signal positions

for the first half frame and

$$\vec{r}_2 = \frac{1}{2}(1 + a_1)\vec{h} + \frac{1}{2}(1 - a_1)\vec{g} + \vec{n}_2 \tag{12.9}$$

for the second half frame, depending on the transmit symbol $a_1 \in \{\pm 1\}$ in the present PPM frame and the transmit symbol $a_0 \in \{\pm 1\}$ in the previous PPM frame. $\vec{h}$ denotes the part of the CIR not causing ISI and $\vec{g}$ is the part of the CIR that causes ISI. Throughout this section we assume that the taps of the CIR are statistically independent normal random variables with zero mean. $\vec{n}_1$ and $\vec{n}_2$ contain the additive white Gaussian noise (AWGN), both with variance σ^2. Since $\vec{r}_1$ and $\vec{r}_2$ are not overlapping, the whole receive signal in the considered PPM frame can be described by

$$\begin{aligned} \vec{r} &= [\vec{r}_1, \vec{r}_2] \\ &= \frac{1}{2}[(1 - a_1)\vec{h}_1 + (1 + a_0)\vec{g}_1 + (1 + a_1)\vec{h}_2 + (1 - a_1)\vec{g}_2] + \vec{n} \end{aligned} \tag{12.10}$$

with

$$\vec{h}_1 = [\vec{h}, 0, \ldots, 0] \qquad\qquad \vec{h}_2 = [0, \ldots, 0, \vec{h}]$$
$$\vec{g}_1 = [\vec{g}, 0, \ldots, 0] \qquad\qquad \vec{g}_2 = [0, \ldots, 0, \vec{g}] \tag{12.11}$$

and

$$\vec{n} = [\vec{n}_1, \vec{n}_2]. \tag{12.12}$$

Both, $\vec{r}_1$ and $\vec{r}_2$ contain $N/2$ elements, i.e., $\vec{r}$ contains N elements.

12.2.2 Symbol-wise Maximum Likelihood Detector

To derive the ML detector, we consider the probability $p(\vec{r}|a_1, a_0, \mathcal{C})$. There, $\mathcal{C}$ denotes the amount of channel state information that is available. As we already proposed in [109], the following cases of CSI are considered for the derivation of the ML detectors: (i) full CSI, (ii) IPDP, (iii) APDP, and (iv) no CSI.

Full Channel State Information

If full CSI is available at the receiver side, i.e., $\mathcal{C}_{\text{full}} = [\vec{h}, \vec{g}]$, only the previous transmit symbol a_0 is unknown. Hence, a_0 has to be averaged out from the desired probability, i.e.,

$$p\left(\vec{r}|a_1 = a_i, \mathcal{C}_{\text{full}}\right) = \int_{-\infty}^{\infty} p(a_0) p\left(\vec{r}|a_1 = a_i, a_0, \vec{h}, \vec{g}\right) da_0. \tag{12.13}$$

Assuming statistically independent noise samples in (12.10),

$$
\begin{aligned}
&p\left(\vec{r}|a_1 = a_i, a_0, \vec{h}, \vec{g}\right) \\
&= p\left(\vec{n} = \vec{r} - \frac{1}{2}(1 - a_1)\vec{h}_1 - \frac{1}{2}(1 + a_0)\vec{g}_1 - \frac{1}{2}(1 + a_1)\vec{h}_2 - \frac{1}{2}(1 - a_1)\vec{g}_2\right) \\
&= \prod_{k=1}^{N} p\left(n_k = r_k - \frac{1}{2}(1 - a_1)h_{1,k} - \frac{1}{2}(1 + a_0)g_{1,k} - \frac{1}{2}(1 + a_1)h_{2,k} - \frac{1}{2}(1 - a_1)g_{2,k}\right).
\end{aligned}
\tag{12.14}
$$

With

$$p\left(n_k\right) = \frac{1}{\sqrt{2\pi\sigma^2}} \cdot \exp\left\{-\frac{n_k^2}{2\sigma^2}\right\} \tag{12.15}$$

the probability density function for the kth factor in (12.14) is given by

$$
\begin{aligned}
&p\left(r_k|a_1 = a_i, a_0, h_{1,k}, h_{2,k}, g_{1,k}, g_{2,k}\right) \\
&= \frac{1}{\sqrt{2\pi\sigma^2}} \cdot \exp\left\{-\frac{1}{2\sigma^2}\left[r_k^2 + \frac{1}{4}(1 - a_1)^2 h_{1,k}^2 + \frac{1}{4}(1 + a_0)^2 g_{1,k}^2 + \frac{1}{4}(1 + a_1)^2 h_{2,k}^2 \right.\right. \\
&+ \frac{1}{4}(1 - a_1)^2 g_{2,k}^2 - (1 - a_1) r_k h_{1,k} - (1 + a_0) r_k g_{1,k} - (1 + a_1) r_k h_{2,k} - (1 - a_1) r_k g_{2,k} \\
&+ \frac{1}{2}(1 - a_1)(1 + a_0) h_{1,k} g_{1,k} + \frac{1}{2}(1 - a_1)(1 - a_1) h_{1,k} g_{2,k} + \frac{1}{2}(1 + a_0)(1 + a_1) h_{2,k} g_{1,k} \\
&\left.\left.+ \frac{1}{2}(1 + a_0)(1 - a_1) g_{1,k} g_{2,k} + \frac{1}{2}(1 + a_1)(1 - a_1) h_{2,k} g_{2,k}\right]\right\}
\end{aligned}
\tag{12.16}
$$

With $h_{1,k} = h_{2,k+N/2} = h_k$ and as $h_{1,k}g_{2,k} = h_{2,k}g_{1,k} = g_{1,k}g_{2,k} = 0$, (12.16) can be simplified such that

$$
p\left(r_k | a_1 = a_i, a_0, h_k, g_k\right)
$$

$$
\propto \exp\left\{-\frac{1}{2\sigma^2}\left[\frac{1}{4}(1-a_1)^2 h_k^2 + \frac{1}{4}(1+a_0)^2 g_k^2 + \frac{1}{4}(1+a_1)^2 h_k^2 + \frac{1}{4}(1-a_1)^2 g_k^2\right.\right.
$$

$$
- (1-a_1)r_k h_k - (1+a_0)r_k g_k - (1+a_1)r_{k+N/2}h_k - (1-a_1)r_{k+N/2}g_k
$$

$$
\left.\left. + \frac{1}{2}(1-a_1)(1+a_0)h_k g_k\right]\right\} \tag{12.17}
$$

In the remainder of this section, we assume equiprobable symbols such that $P[a_1 = 1] = P[a_1 = -1]$ and $p(a_0) = \frac{1}{2}\delta(a_0 - 1) + \frac{1}{2}\delta(a_0 + 1)$. Thus,

$$
p\left(r_k | a_1 = a_i, h_k, g_k\right)
$$

$$
\propto \frac{1}{2}\exp\left\{-\frac{1}{2\sigma^2}\left[\frac{1}{4}(1-a_1)^2 h_k^2 + g_k^2 + \frac{1}{4}(1+a_1)^2 h_k^2 + \frac{1}{4}(1-a_1)^2 g_k^2\right.\right.
$$

$$
\left.\left. - (1-a_1)r_k h_k - 2r_k g_k - (1+a_1)r_{k+N/2}h_k - (1-a_1)r_{k+N/2}g_k + (1-a_1)h_k g_k\right]\right\}
$$

$$
+ \frac{1}{2}\exp\left\{-\frac{1}{2\sigma^2}\left[\frac{1}{4}(1-a_1)^2 h_k^2 + \frac{1}{4}(1+a_1)^2 h_k^2 + \frac{1}{4}(1-a_1)^2 g_k^2\right.\right.
$$

$$
\left.\left. - (1-a_1)r_k h_k - (1+a_1)r_{k+N/2}h_k - (1-a_1)r_{k+N/2}g_k\right]\right\}
$$

$$
= \frac{1}{2}\exp\left\{-\frac{1}{2\sigma^2}\left[\frac{1}{4}(1+a_1)^2 h_k^2 + \frac{1}{4}(1-a_1)^2 h_k^2 + \frac{1}{4}(1-a_1)^2 g_k^2\right.\right.
$$

$$
\left.\left. - (1+a_1)r_{k+N/2}h_k - (1-a_1)r_{k+N/2}g_k - (1-a_1)r_k h_k\right]\right\}
$$

$$
\cdot \left(\exp\left\{-\frac{1}{2\sigma^2}\left[(1-a_1)h_k g_k + g_k^2 - 2r_k g_k\right]\right\} + 1\right). \tag{12.18}
$$

With the definition of the log-likelihood ratio

$$
L - \ln\left(\frac{p\left(\vec{r}|a_1 = 1, \mathcal{C}\right)}{p\left(r|a_1 = -1, \mathcal{C}\right)}\right) \tag{12.19}
$$

we obtain in the case of full CSI

$$
L = \frac{1}{2\sigma^2}\sum_{k=1}^{N/2}\left(2r_{k+N/2}h_k - 2r_k h_k + g_k^2 - 2r_{k+N/2}g_k\right)
$$

$$
- \ln\left[\frac{\exp\left\{-\sum_{k=1}^{N/2}\frac{h_k g_k}{\sigma^2}\right\}\exp\left\{-\sum_{k=1}^{N/2}\frac{g_k^2 - 2r_k g_k}{2\sigma^2}\right\} + 1}{\exp\left\{-\sum_{k=1}^{N/2}\frac{g_k^2 - 2r_k g_k}{2\sigma^2}\right\} + 1}\right]. \tag{12.20}
$$

We refer to this receiver in the following as $\mathrm{ML}_{\text{full,ISI}}$. If no ISI is present, the log-likelihood rule yields the matched filter receiver, i.e.,

$$L = \frac{1}{\sigma^2} \sum_{k=1}^{N/2} \left(r_{k+N/2} h_k - r_k h_k \right), \tag{12.21}$$

where the receive signal is correlated with a template and thus the channel taps are coherently combined. This receiver is called in the remainder $\mathrm{ML}_{\text{full}}$.

Instantaneous Power Delay Profile

In a next step, we assume that only the IPDP is available at the receiver side. With this assumption we can write $\vec{h} = \vec{x} \odot \vec{z}$ and $\vec{g} = \vec{u} \odot \vec{v}$, where $\vec{x}$ and $\vec{u}$ denote the magnitudes of the channel impulse response (i.e. the IPDP) and $\vec{z}$ and $\vec{v}$ denote the signs of $\vec{h}$ and $\vec{g}$, respectively. Hence, the CSI is given by $\mathcal{C}_{\text{IPDP}} = [\vec{x}, \vec{u}]$. Since the signs z_k and v_k are equiprobable, their probability density functions are given by

$$\begin{aligned}
p(z_k) &= \frac{1}{2}\delta(z_k - 1) + \frac{1}{2}\delta(z_k + 1) \\
p(v_k) &= \frac{1}{2}\delta(v_k - 1) + \frac{1}{2}\delta(v_k + 1).
\end{aligned} \tag{12.22}$$

The signs of the CIRs are unknown at the receiver side and have to be averaged out, so that the desired probability depends only on the transmit symbol and the IPDP. This yields

$$\begin{aligned}
&p\left(\vec{r}|a_1 = a_i, \mathcal{C}_{\text{IPDP}}\right) \\
&= \prod_{k=1}^{N} \int_{-\infty}^{\infty} \int_{-\infty}^{\infty} \int_{-\infty}^{\infty} p(a_0)p(z_k)p(v_k)p\left(r_k|a_1 = a_i, a_0, x_k, z_k, u_k, v_k\right) dv_k dz_k da_0.
\end{aligned} \tag{12.23}$$

With (12.16),

$$\begin{aligned}
&p\left(r_k|a_1 = a_i, a_0, x_k, z_k, u_k, v_k\right) \\
&\propto \exp\left\{ -\frac{1}{2\sigma^2} \left[(x_k z_k)^2 + \frac{1}{4}\left((1+a_0)^2 + (1-a_1)^2\right)(u_k v_k)^2 \right.\right. \\
&\quad - \left((1+a_0)r_k + (1-a_1)r_{k+N/2}\right)(u_k v_k) \\
&\quad \left.\left. - \left((1-a_1)r_k + (1+a_1)r_{k+N/2} - \frac{1}{2}(1-a_1)(1+a_0)(u_k v_k)\right)(x_k z_k)\right]\right\}.
\end{aligned} \tag{12.24}$$

Averaging over v_k, z_k and a_0 yields

$$p\left(r_k|a_1 = a_i, x_k, u_k\right)$$

$$\propto \frac{1}{8}\exp\left\{-\frac{1}{2\sigma^2}\left[x_k^2 + \frac{1}{4}\left(4 + (1 - a_1)^2\right)u_k^2 - \left(2r_k + (1 - a_1)r_{k+N/2}\right)u_k\right.\right.$$

$$+ \left.\left.\left(-(1 - a_1)r_k - (1 + a_1)r_{k+N/2} + (1 - a_1)u_k\right)x_k\right]\right\}$$

$$+ \frac{1}{8}\exp\left\{-\frac{1}{2\sigma^2}\left[x_k^2 + \frac{1}{4}\left(4 + (1 - a_1)^2\right)u_k^2 + \left(2r_k + (1 - a_1)r_{k+N/2}\right)u_k\right.\right.$$

$$+ \left.\left.\left(-(1 - a_1)r_k - (1 + a_1)r_{k+N/2} - (1 - a_1)u_k\right)x_k\right]\right\}$$

$$+ \frac{1}{8}\exp\left\{-\frac{1}{2\sigma^2}\left[x_k^2 + \frac{1}{4}\left(4 + (1 - a_1)^2\right)u_k^2 - \left(2r_k + (1 - a_1)r_{k+N/2}\right)u_k\right.\right.$$

$$- \left.\left.\left(-(1 - a_1)r_k - (1 + a_1)r_{k+N/2} + (1 - a_1)u_k\right)x_k\right]\right\}$$

$$+ \frac{1}{8}\exp\left\{-\frac{1}{2\sigma^2}\left[x_k^2 + \frac{1}{4}\left(4 + (1 - a_1)^2\right)u_k^2 + \left(2r_k + (1 - a_1)r_{k+N/2}\right)u_k\right.\right.$$

$$- \left.\left.\left(-(1 - a_1)r_k - (1 + a_1)r_{k+N/2} - (1 - a_1)u_k\right)x_k\right]\right\}$$

$$+ \frac{1}{8}\exp\left\{-\frac{1}{2\sigma^2}\left[x_k^2 + \frac{1}{4}(1 - a_1)^2 u_k^2 - (1 - a_1)r_{k+N/2}u_k\right.\right.$$

$$+ \left.\left.\left(-(1 - a_1)r_k - (1 + a_1)r_{k+N/2}\right)x_k\right]\right\}$$

$$+ \frac{1}{8}\exp\left\{-\frac{1}{2\sigma^2}\left[x_k^2 + \frac{1}{4}(1 - a_1)^2 u_k^2 + (1 - a_1)r_{k+N/2}u_k\right.\right.$$

$$+ \left.\left.\left(-(1 - a_1)r_k - (1 + a_1)r_{k+N/2}\right)x_k\right]\right\}$$

$$+ \frac{1}{8}\exp\left\{-\frac{1}{2\sigma^2}\left[x_k^2 + \frac{1}{4}(1 - a_1)^2 u_k^2 - (1 - a_1)r_{k+N/2}u_k\right.\right.$$

$$- \left.\left.\left(-(1 - a_1)r_k - (1 + a_1)r_{k+N/2}\right)x_k\right]\right\}$$

$$+ \frac{1}{8}\exp\left\{-\frac{1}{2\sigma^2}\left[x_k^2 + \frac{1}{4}(1 - a_1)^2 u_k^2 + (1 - a_1)r_{k+N/2}u_k\right.\right.$$

$$- \left.\left.\left(-(1 - a_1)r_k - (1 + a_1)r_{k+N/2}\right)x_k\right]\right\} \tag{12.25}$$

Inserting the obtained results from (12.25) into (12.19) we get for the log-likelihood ratio assuming the knowledge of the IPDP,

$$L = \sum_{k=1}^{N/2} \left(\ln \left[\exp \left\{ -\frac{u_k^2}{2\sigma^2} \right\} \left(\cosh \left(\frac{r_{k+N/2}x_k + r_k u_k}{\sigma^2} \right) + \cosh \left(\frac{r_{k+N/2}x_k - r_k u_k}{\sigma^2} \right) \right) \right. \right.$$
$$\left. \left. + 2 \cosh \left(\frac{r_{k+N/2}x_k}{\sigma^2} \right) \right] \right)$$
$$- \sum_{k=1}^{N/2} \left(\ln \left[\exp \left\{ -\frac{u_k^2}{2\sigma^2} \right\} \left(\cosh \left(\frac{r_k x_k + r_{k+N/2}u_k}{\sigma^2} \right) + \cosh \left(\frac{r_k x_k - r_{k+N/2}u_k}{\sigma^2} \right) \right) \right. \right.$$
$$+ \exp \left\{ -\frac{u_k^2 - r_k x_k}{\sigma^2} \right\} \cosh \left(\frac{(r_{k+N/2} + r_k - x_k)u_k}{\sigma^2} \right)$$
$$\left. \left. + \exp \left\{ -\frac{u_k^2 + r_k x_k}{\sigma^2} \right\} \cosh \left(\frac{(r_{k+N/2} + r_k + x_k)u_k}{\sigma^2} \right) \right] \right). \tag{12.26}$$

Without ISI (i.e. $\vec{u} = 0$), the log-likelihood ratio (12.26) reduces to

$$L = \sum_{k=1}^{N/2} \left(\ln \left[\cosh \left(\frac{r_{k+N/2}x_k}{\sigma^2} \right) \right] - \ln \left[\cosh \left(\frac{r_k x_k}{\sigma^2} \right) \right] \right), \tag{12.27}$$

i.e., the receive signal is weighted with the magnitude of the corresponding channel tap. The receiver, which considers the ISI, is denoted in the remainder as $\text{ML}_{\text{IPDP,ISI}}$ while the latter one is referred to as ML_{IPDP}.

Average Power Delay Profile

In this case, the receiver knows the correlation matrices of $\vec{h}$ and $\vec{g}$, i.e.,

$$\Lambda_{hh} = \begin{bmatrix} \mathcal{E}[h_1^2] & \cdots & 0 \\ \vdots & \ddots & \vdots \\ 0 & \cdots & \mathcal{E}[h_{N/2}^2] \end{bmatrix} = \begin{bmatrix} \lambda_{h,1} & \cdots & 0 \\ \vdots & \ddots & \vdots \\ 0 & \cdots & \lambda_{h,N/2} \end{bmatrix} \tag{12.28}$$

and

$$\Lambda_{gg} = \begin{bmatrix} \mathcal{E}[g_1^2] & \cdots & 0 \\ \vdots & \ddots & \vdots \\ 0 & \cdots & \mathcal{E}[g_{N/2}^2] \end{bmatrix} = \begin{bmatrix} \lambda_{g,1} & \cdots & 0 \\ \vdots & \ddots & \vdots \\ 0 & \cdots & \lambda_{g,N/2} \end{bmatrix}, \tag{12.29}$$

respectively. We rearrange the receive vector such that the kth tap of $\vec{r}_1$ and $\vec{r}_2$ are adjacent, i.e.,

$$\vec{r} = \left[r_1, r_{N/2+1}, r_2, r_{N/2+2}, \ldots, r_{N/2}, r_N \right]. \tag{12.30}$$

Hence, the correlation matrix Λ_{rr} can be written as

$$\Lambda_{rr} = \begin{bmatrix} \lambda_{r,11} & \lambda_{r,12} & \cdots & 0 & 0 \\ \lambda_{r,21} & \lambda_{r,22} & \cdots & 0 & 0 \\ \vdots & \vdots & \ddots & \vdots & \vdots \\ 0 & 0 & \cdots & \lambda_{r,N-1N-1} & \lambda_{r,N-1N} \\ 0 & 0 & \cdots & \lambda_{r,NN-1} & \lambda_{r,NN} \end{bmatrix} \tag{12.31}$$

with

$$\lambda_{r,kk} = (1 - a_1)^2 \lambda_{h,\lceil k/2 \rceil} + (1 + a_0)^2 \lambda_{g,\lceil k/2 \rceil} + \sigma^2 \tag{12.32}$$

for odd k's and

$$\lambda_{r,kk} = (1 + a_1)^2 \lambda_{h,\lceil k/2 \rceil} + (1 - a_1)^2 \lambda_{g,\lceil k/2 \rceil} + \sigma^2 \tag{12.33}$$

for even k's. $\lceil \cdot \rceil$ denotes a rounding up to the next higher integer number. The correlated elements on the secondary diagonals are given by

$$\lambda_{r,kk+1} = \lambda_{r,k+1k} = (1 + a_0)(1 - a_1)\lambda_{g,\lceil k/2 \rceil}. \tag{12.34}$$

According to [111], the desired probability assuming the knowledge of the correlation matrix Λ_{rr} is defined as

$$p\left(\vec{r}|a_1 = a_i, a_0, \Lambda_{rr}\right) = \left(\frac{1}{\sqrt{2\pi}}\right)^N \frac{1}{\sqrt{\Delta_{rr}}} \exp\left(-\frac{1}{2}\vec{r}^T \Lambda_{rr}^{-1} \vec{r}\right), \tag{12.35}$$

where Δ_{rr} denotes the determinant of Λ_{rr} and $\vec{r}^T$ is the transpose of $\vec{r}$. The determinant Δ_{rr} can be obtained as

$$\Delta_{rr} = \left(\frac{1}{16}\right)^{\frac{N}{2}} \prod_{k=1}^{N/2} \left[16\sigma^4 + \lambda_{h,k}\lambda_{g,k}\left((a_0 + 1)^2(a_1 + 1)^2 + (a_1 - 1)^4\right)\right.$$
$$\left. + 4\lambda_{g,k}\sigma^2\left((a_0 + 1)^2 + (a_1 - 1)^2\right) + 4\lambda_{h,k}\sigma^2\left((a_1 - 1)^2 + (a_1 + 1^2)\right)\right]. \tag{12.36}$$

The inverse of Λ_{rr} has the same structure as the original correlation matrix (12.31) with the elements

$$(\Lambda_{rr})_{kk}^{-1} = \frac{4(\lambda_{g,\lceil k/2 \rceil}(a_1 - 1)^2 + \lambda_{h,\lceil k/2 \rceil}(a_1 + 1)^2 + 4\sigma^2)}{\psi_{\lceil k/2 \rceil}} \tag{12.37}$$

for odd k's and the elements

$$(\Lambda_{rr})_{kk}^{-1} = \frac{4(\lambda_{g,\lceil k/2\rceil}(a_0 + 1)^2 + \lambda_{h,\lceil k/2\rceil}(a_1 - 1)^2 + 4\sigma^2)}{\psi_{\lceil k/2\rceil}} \tag{12.38}$$

for even k's. The elements on the secondary diagonal of the inverse can be derived as

$$(\Lambda_{rr})_{kk+1}^{-1} = \frac{4\lambda_{g,\lceil k/2\rceil}(a_0 + 1)(a_1 - 1)}{\psi_{\lceil k/2\rceil}}. \tag{12.39}$$

The denominator $\psi_{\lceil k/2\rceil}$ in (12.37) - (12.39) is given by

$$\psi_{\lceil k/2\rceil} = \lambda_{g,\lceil k/2\rceil}\lambda_{h,\lceil k/2\rceil}(2 + 2a_0(1 + a_1)^2 + a_0^2(1 + a_1)^2 + a_1(-2 + a_1(7 + (a_1 - 4)a_1)))$$

$$+ 4\lambda_{g,\lceil k/2\rceil}(2 + a_0(2 + a_0) + (a_1 - 2)a_1)\sigma^2 + 16\lambda_{h,\lceil k/2\rceil}\sigma^2 + 16\sigma^4. \tag{12.40}$$

Evaluating (12.35) and averaging over a_0, the log-likelihood ratio for $\text{ML}_{\text{APDP,ISI}}$ is obtained as

$$L = \sum_{k=1}^{N/2} \ln\left(\frac{1}{\sqrt{\zeta_k}} \exp\left\{-\frac{r_k^2\alpha_k + r_{k+N/2}^2\beta_k}{2\zeta_k}\right\} + \frac{1}{\sqrt{\sigma^2\alpha_k}} \exp\left\{-\frac{1}{2}\left(\frac{r_k^2}{\sigma^2} + \frac{r_{k+N/2}^2}{\alpha_k}\right)\right\}\right)$$

$$- \sum_{k=1}^{N/2} \ln\left(\frac{1}{\sqrt{\zeta_k}} \exp\left\{-\frac{r_k^2\beta_k + r_{k+N/2}^2\alpha_k}{2\zeta_k}\right\}\right.$$

$$\left. + \frac{1}{\sqrt{\zeta_k + \lambda_{g,k}\sigma^2}} \cdot \exp\left\{-\frac{r_k^2\beta_k + r_{k+N/2}^2(\beta_k + \lambda_{h,k}) - 2r_k r_{k+N/2}\lambda_{g,k}}{2(\zeta_k + \lambda_{g,k}\sigma^2)}\right\}\right) \tag{12.41}$$

with $\alpha_k = \lambda_{h,k} + \sigma^2$, $\beta_k = \lambda_{g,k} + \sigma^2$, and $\zeta_k = \alpha_k\beta_k$.

In the case where no ISI is present, i.e., $\lambda_{g,k} = 0$, we get the ML_{APDP}, where the log-likelihood ratio is given by

$$L = \frac{1}{2\sigma^2} \sum_{k=1}^{N/2} \frac{r_{k+N/2}^2 - r_k^2}{1 + \frac{\sigma^2}{\lambda_{h,k}}}, \tag{12.42}$$

i.e., this is a typical energy detector, whose output is weighted with the APDP. Channel taps that are in average small are considered for decision less than large ones. The log-likelihood ratio without ISI corresponds to the result presented in [112].

No Channel State Information

Finally, we consider the case that the receiver only knows the average energy of the CIR and the noise variance σ^2. Such a receiver can be regarded as a special case of the receiver with APDP knowledge. Since the channel taps are not known, all diagonal elements of (12.28) and (12.29) are assumed to be equal, i.e., $\forall k : \lambda_{h,k} = \lambda_{g,k} = \lambda'$. As $\alpha_k = \beta_k = \alpha' = \lambda' + \sigma^2$ and $\zeta_k = \alpha'^2$, the result in (12.41) changes to

$$
\begin{aligned}
L = &\sum_{k=1}^{N/2} \ln\left(\frac{1}{\lambda' + \sigma^2} \exp\left\{-\frac{r_k^2 + r_{k+N/2}^2}{2(\lambda' + \sigma^2)}\right\}\right. \\
&+ \frac{1}{\sqrt{\sigma^2(\lambda' + \sigma^2)}} \exp\left\{-\frac{1}{2}\left(\frac{r_k^2}{\sigma^2} + \frac{r_{k+N/2}^2}{\lambda' + \sigma^2}\right)\right\}\right) \\
&- \sum_{k=1}^{N/2} \ln\left(\frac{1}{\lambda' + \sigma^2} \exp\left\{-\frac{r_k^2 + r_{k+N/2}^2}{2(\lambda' + \sigma^2)}\right\}\right. \\
&+ \frac{1}{\sqrt{(\lambda' + \sigma^2)^2 + \lambda'\sigma^2}} \cdot \exp\left\{-\frac{r_k^2(\lambda' + \sigma^2) + r_{k+N/2}^2(2\lambda' + \sigma^2) - 2r_k r_{k+N/2}\lambda'}{2\left((\lambda' + \sigma^2)^2 + \lambda'\sigma^2\right)}\right\}\right).
\end{aligned}
$$
(12.43)

For the no ISI case, inserting $\lambda_{h,k} = \lambda'$ into (12.42) yields the common energy detector

$$
\begin{aligned}
L =& \frac{1}{2\sigma^2(1 + \frac{\sigma^2}{\lambda'})} \sum_{k=1}^{N/2} r_{k+N/2}^2 - r_k^2 \\
&\propto \sum_{k=1}^{N/2} r_{k+N/2}^2 - r_k^2,
\end{aligned}
$$
(12.44)

which is referred to as ED in the following.

12.2.3 Performance Results

To see the impact of the CSI we compare the performance of the above derived ML receivers by means of bit error rate (BER) simulations. The BER curves are plotted over the signal-to-noise ratio E_b/N_0, where E_b denotes the energy per bit and $N_0/2$ is the noise power spectral density. We assume uniformly distributed channel taps within the duration of the channel impulse response. To achieve a data rate of 50 Mbps with 2PPM one bit has to be transmitted every 20 ns. Hence, one PPM frame has a duration of $T = 20$ ns, i.e., one PPM slot has a duration of 10 ns and ISI occurs for channel impulse responses with a duration of more than 10 ns. We consider the cases that no ISI, weak ISI, and strong ISI occur. In case of no ISI the CIR has a duration of 10 ns, in case of weak ISI 14 ns, and in case of strong ISI 17 ns.

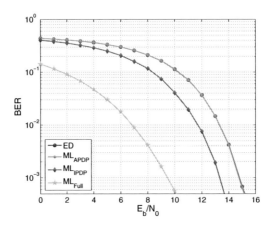

Figure 12.2: Bit error rate curves in the case of no ISI

No ISI The BER curves in case of no ISI are shown in Fig. 12.2. As expected, the higher the amount of CSI the better is the performance of the corresponding receiver. However, this performance improvement is achieved at the cost of higher receiver complexity. The ML_{full}, which does a coherent combining, performs the best and meets the matched filter (MF) bound. Since the ML_{IPDP} has no information on the signs of the channel taps, no coherent combining is possible. The performance of this receiver structure is about 5 dB worse compared to the MF. A further performance degradation of about 2 dB is observed for the ML_{APDP}. There, the channel taps cannot be weighted with the instantaneous power but only with the average power. Since we assume uniformly distributed channel taps and a CIR duration of 10 ns, the ML_{APDP} equals the energy detector in this case. Therefore, the performance of these both receivers is the same.

Weak ISI The performance of the ML receivers not considering the ISI becomes worse in the presence of ISI as it can be seen in Fig. 12.3. As expected, the performance improves if the ML receivers are adapted to the ISI. The ML_{full} is about 2 dB worse compared with the no ISI case. However, the $\text{ML}_{\text{full,ISI}}$ loses only about 0.8 dB. A similar observation can be made for the

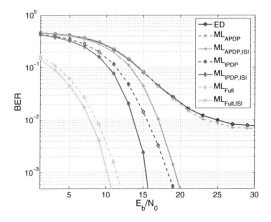

Figure 12.3: Bit error rate curves in the case of weak ISI

ML receivers with IPDP. While the ML_{IPDP} degrades by about 4 dB, the $ML_{IPDP,ISI}$ is only about 1.5 dB worse compared to the result in Fig. 12.2. Therefore, the MLs with full CSI and IPDP do not necessarily have to be optimized for ISI if only weak ISI is present. For the ML_{APDP} and for the energy detector a different behavior can be observed. Due to the ISI, the influence of the noise on the BER performance is not dominant for high E_b/N_0 values and the BER curves approach an error floor. The $ML_{APDP,ISI}$ shows a much better performance and is about 4 dB worse compared to the no ISI case. Although the receiver complexity of the $ML_{APDP,ISI}$ is somewhat higher than the one of the ML_{APDP}, the $ML_{APDP,ISI}$ should be used in case of ISI due to the much better performance.

Strong ISI The BER curves for strong ISI are plotted in Fig. 12.4. As described above, we also compare the ML receivers that are adapted to the ISI with the ones that are optimized for no ISI. The ML receivers not optimized for ISI perform in the strong ISI scenario much worse than in the weak ISI scenario except for the ML_{full}. The performance of the ML_{full} degrades only slightly while the energy detector and the ML_{APDP} exhibit an error floor at about $8 \cdot 10^{-2}$. An error floor for high E_b/N_0 values can also be expected for the ML_{IPDP}. The ML receivers

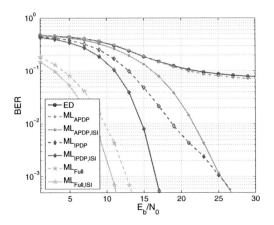

Figure 12.4: Bit error rate curves in the case of strong ISI

that consider ISI show a much better performance. The $ML_{full,ISI}$ is only about 1.6 dB, the $ML_{IPDP,ISI}$ about 2.6 dB, and the $ML_{APDP,ISI}$ about 6 dB worse than the corresponding "no ISI" BER curves at a BER $= 10^{-3}$. For the $ML_{full,ISI}$ and the $ML_{IPDP,ISI}$ a precise channel estimation is necessary, which increases the receiver complexity. Hence, these both structures are rather suited for applications with relaxed complexity constraints or applications requiring very high data rates, where strong ISI occurs.

Conclusions   We presented a family of symbol-wise detectors for a binary PPM, which utilize partial channel state information to improve the robustness to ISI. We considered the cases of no CSI, APDP, IPDP, and full CSI. In the presence of ISI the performance of the ML receivers that are not adapted to ISI degraded and approached an error floor for high signal-to-noise ratios. If the ISI is considered by the ML, the performance can be improved substantially. However, from the equations and the BER curves a tradeoff between complexity and performance can be observed. The performance becomes better with increasing amount of CSI and by considering the ISI but it results in much higher receiver complexity. Moreover, the ML receivers with partial CSI require an estimation of the CSI, which increases the complexity even more.

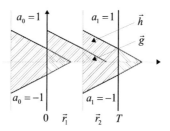

Figure 12.5: Schematic of the considered receive signal positions

Hence, for a very simple UWB BAN the ED receiver is the most promising solution, although the performance is worse compared to the receivers with partial CSI.

12.3 Maximum Likelihood Receivers for Transmitted Reference Systems in the Presence of ISI

12.3.1 System Model

In this section, transmitted reference systems in the presence of ISI are investigated. Similar to the derivation in section 12.2.1, it is assumed in the remainder of this section that the maximum duration of the channel impulse response is smaller than the duration of one time frame T, i.e., the ISI has only an impact on the next TR half symbol. Time-hopping is omitted again for the purpose of better clarity of the derivation but can be easily included. It is assumed that the reference signal is always positive while the sign of the data pulse is determined by the transmit symbol as depicted in Fig. 12.5. The sampled receive signal in the considered time frame is given by

$$\vec{r}_1 = \vec{h} + a_0\vec{g} + \vec{n}_1 \tag{12.45}$$

for the first half frame and

$$\vec{r}_2 = a_1\vec{h} + \vec{g} + \vec{n}_2 \tag{12.46}$$

for the second half frame, depending on the transmit symbol $a_1 \in \{\pm 1\}$ in the present frame and the transmit symbol $a_0 \in \{\pm 1\}$ in the previous frame. $\vec{h}$ denotes the part of the CIR not causing ISI and $\vec{g}$ is the part of the CIR that causes ISI. Throughout this section it is assumed that the taps of the CIR are statistically independent normal random variables with zero mean. $\vec{n}_1$ and $\vec{n}_2$ contain the additive white Gaussian noise (AWGN), both with variance σ^2. As well as in section 12.3.1 the whole receive signal in the considered frame can be described by

$$
\begin{aligned}
\vec{r} &= [\vec{r}_1, \vec{r}_2] \\
&= \vec{h}_1 + a_0 \vec{g}_1 + a_1 \vec{h}_2 + \vec{g}_2 + \vec{n}
\end{aligned}
\tag{12.47}
$$

with

$$
\begin{aligned}
\vec{h}_1 &= [\vec{h}, 0, \ldots, 0] & \vec{h}_2 &= [0, \ldots, 0, \vec{h}] \\
\vec{g}_1 &= [\vec{g}, 0, \ldots, 0] & \vec{g}_2 &= [0, \ldots, 0, \vec{g}]
\end{aligned}
\tag{12.48}
$$

and

$$
\vec{n} = [\vec{n}_1, \vec{n}_2].
\tag{12.49}
$$

Both, $\vec{r}_1$ and $\vec{r}_2$ contain $N/2$ elements, i.e., $\vec{r}$ contains N elements. The output of the classical TR cross-correlation receiver structure [22] is given by

$$
y = \vec{r}_1^T \vec{r}_2 = \sum_{k=1}^{N/2} r_k \cdot r_{k+N/2}.
\tag{12.50}
$$

12.3.2 Symbol-wise Maximum Likelihood Detector

As in section 12.2.2, the probability $p(\vec{r}|a_1, a_0, \mathcal{C})$ is considered to derive the ML detector. Again, $\mathcal{C}$ denotes the level of channel state information that is available and the following cases of CSI are considered for the derivation of the ML detectors: (i) full CSI, (ii) IPDP, (iii) APDP, and (iv) no CSI.

Full Channel State Information

In the case of full CSI at the receiver side, i.e., $\mathcal{C}_{\text{full}} = [\vec{h}, \vec{g}]$, only the previous transmit symbol a_0 is unknown. Hence, a_0 has to be averaged out from the desired likelihood, i.e.,

$$p\left(\vec{r}|a_1 = a_i, \mathcal{C}_{\text{full}}\right) = \int_{-\infty}^{\infty} p(a_0)p\left(\vec{r}|a_1 = a_i, a_0, \vec{h}, \vec{g}\right) da_0. \tag{12.51}$$

Assuming statistically independent noise samples in (12.10)

$$
\begin{aligned}
p\left(\vec{r}|a_1 = a_i, a_0, \vec{h}, \vec{g}\right) &= p\left(\vec{n} = \vec{r} - \vec{h}_1 - a_0\vec{g}_1 - a_1\vec{h}_2 - \vec{g}_2\right) \\
&= \prod_{k=1}^{N}\left(n_k = r_k - h_{1,k} - a_0 g_{1,k} - a_1 h_{2,k} - g_{2,k}\right).
\end{aligned}
\tag{12.52}
$$

With

$$p\left(n_k\right) = \frac{1}{\sqrt{2\pi\sigma^2}} \cdot \exp\left\{-\frac{n_k^2}{2\sigma^2}\right\} \tag{12.53}$$

the probability density function for the kth in (12.52) is given by

$$
\begin{aligned}
&p\left(r_k|a_1 = a_i, a_0, h_{1,k}, h_{2,k}, g_{1,k}, g_{2,k}\right) \\
&= \frac{1}{\sqrt{2\pi\sigma^2}} \cdot \exp\left\{-\frac{1}{2\sigma^2}\left[r_k^2 + h_{1,k}^2 + g_{1,k}^2 + h_{2,k}^2 + g_{2,k}^2\right.\right. \\
&\left.\left. - 2r_k h_{1,k} - 2a_0 r_k g_{1,k} - 2a_1 r_k h_{2,k} - 2r_k g_{2,k} + 2a_0 h_{1,k} g_{1,k} + 2a_1 h_{2,k} g_{2,k}\right]\right\}
\end{aligned}
\tag{12.54}
$$

With $h_{1,k} = h_{2,k+N/2} = h_k$ and $p(a_0) = \frac{1}{2}\delta(a_0 - 1) + \frac{1}{2}\delta(a_0 + 1)$ the terms relevant for the likelihood ratio in (12.54) yield

$$
\begin{aligned}
&p\left(r_k|a_1 = a_i, h_k, g_k\right) \\
&\propto \frac{1}{2} \exp\left\{-\frac{1}{2\sigma^2}\left[-2r_k h_k - 2r_{k+N/2} h_k a_1 - 2r_k g_k - 2r_{k+N/2} g_k + 2h_k g_k + 2h_k g_k a_1\right]\right\} \\
&+ \frac{1}{2} \exp\left\{-\frac{1}{2\sigma^2}\left[-2r_k h_k - 2r_{k+N/2} h_k a_1 + 2r_k g_k - 2r_{k+N/2} g_k - 2h_k g_k + 2h_k g_k a_1\right]\right\}
\end{aligned}
\tag{12.55}
$$

As in section 12.2.2, it is assumed in the remainder of this section that $P[a_1 = 1] = P[a_1 = -1]$. Thus, the log-likelihood ratio is defined as

$$L = \ln\left(\frac{p\left(\vec{r}|a_1 = 1, \mathcal{C}\right)}{p\left(\vec{r}|a_1 = -1, \mathcal{C}\right)}\right) \tag{12.56}$$

and one obtains in the case of full CSI

$$L = \frac{1}{\sigma^2} \sum_{k=1}^{N/2} \left(2r_{k+N/2}h_k\right) - \ln \left[\frac{\exp\left\{-\sum_{k=1}^{N/2} \frac{r_k g_k}{\sigma^2}\right\} + \exp\left\{\sum_{k=1}^{N/2} \frac{r_k g_k - 2h_k g_k}{\sigma^2}\right\}}{\exp\left\{\sum_{k=1}^{N/2} \frac{r_k g_k}{\sigma^2}\right\} + \exp\left\{-\sum_{k=1}^{N/2} \frac{r_k g_k - 2h_k g_k}{\sigma^2}\right\}} \right].$$

(12.57)

This receiver is referred to in the following as $ML_{TR,full,ISI}$. If no ISI is present, the log-likelihood rule yields the matched filter receiver, i.e.,

$$L = \frac{1}{\sigma^2} \sum_{k=1}^{N/2} \left(2r_{k+N/2}h_k\right).$$

(12.58)

This receiver is called in the remainder $ML_{TR,full}$. The log-likelihood rule in (12.58) corresponds to a "classical" matched filter receiver, where only the data pulse is considered. The reference pulse, which is usually used for implicit channel estimation and carries no data information, is not used for the log-likelihood estimation with full CSI. Therefore, a performance degradation compared to 2PAM is expected due to the transmission of the second pulse that is not required for decision.

Instantaneous Power Delay Profile

In a next step, only the knowledge of the IPDP is assumed at the receiver side. As in section 12.2.2 one can write $\vec{h} = \vec{x} \odot \vec{z}$ and $\vec{g} = \vec{u} \odot \vec{v}$, where $\vec{x}$ and $\vec{u}$ denote the magnitudes of the channel impulse response (i.e. the IPDP) and $\vec{z}$ and $\vec{v}$ denote the signs of $\vec{h}$ and $\vec{g}$, respectively. Hence, the CSI is given by $C_{IPDP} = [\vec{x}, \vec{u}]$. Since the signs z_k and v_k are equiprobable, the probability density functions are given by

$$p(z_k) = \frac{1}{2}\delta(z_k - 1) + \frac{1}{2}\delta(z_k + 1)$$
$$p(v_k) = \frac{1}{2}\delta(v_k - 1) + \frac{1}{2}\delta(v_k + 1).$$

(12.59)

The signs of the CIRs are unknown at the receiver side and have to be averaged out so that the desired probability depends only on the transmit symbol and the IPDP. This yields

$$p\left(\vec{r}|a_1 = a_i, \mathcal{C}_{\text{IPDP}}\right)$$

$$= \prod_{k=1}^{N} \int_{-\infty}^{\infty} \int_{-\infty}^{\infty} \int_{-\infty}^{\infty} p(a_0)p(z_k)p(v_k)p\left(r_k|a_1 = a_i, a_0, x_k, z_k, u_k, v_k\right) dv_k dz_k da_0. \quad (12.60)$$

Using (12.55),

$$p\left(r_k|a_1 = a_i, a_0, x_k, z_k, u_k, v_k\right) \propto \exp\left\{ -\frac{1}{\sigma^2} \left[(x_k z_k)^2 + (u_k v_k)^2 \right.\right.$$

$$\left.\left. + \left(a_0 r_k + r_{k+N/2} - h_k a_0 - h_k a_1\right)(u_k v_k) - \left(r_k + a_1 r_{k+N/2}\right)(x_k z_k) \right] \right\}. \quad (12.61)$$

Averaging over v_k, z_k and a_0 yields

$$p\left(r_k|a_1 = a_i, x_k, u_k\right)$$

$$\propto \frac{1}{8} \exp\left\{ -\frac{-(r_k + r_{k+N/2} - x_k - a_1 x_k)u_k - (r_k + a_1 r_{k+N/2})x_k}{\sigma^2} \right\}$$

$$+ \frac{1}{8} \exp\left\{ -\frac{(r_k + r_{k+N/2} - x_k - a_1 x_k)u_k - (r_k + a_1 r_{k+N/2})x_k}{\sigma^2} \right\}$$

$$+ \frac{1}{8} \exp\left\{ -\frac{-(r_k + r_{k+N/2} - x_k - a_1 x_k)u_k + (r_k + a_1 r_{k+N/2})x_k}{\sigma^2} \right\}$$

$$+ \frac{1}{8} \exp\left\{ -\frac{(r_k + r_{k+N/2} - x_k - a_1 x_k)u_k + (r_k + a_1 r_{k+N/2})x_k}{\sigma^2} \right\}$$

$$+ \frac{1}{8} \exp\left\{ -\frac{-(-r_k + r_{k+N/2} + x_k - a_1 x_k)u_k - (r_k + a_1 r_{k+N/2})x_k}{\sigma^2} \right\}$$

$$+ \frac{1}{8} \exp\left\{ -\frac{(-r_k + r_{k+N/2} + x_k - a_1 x_k)u_k - (r_k + a_1 r_{k+N/2})x_k}{\sigma^2} \right\}$$

$$+ \frac{1}{8} \exp\left\{ -\frac{-(-r_k + r_{k+N/2} + x_k - a_1 x_k)u_k + (r_k + a_1 r_{k+N/2})x_k}{\sigma^2} \right\}$$

$$+ \frac{1}{8} \exp\left\{ -\frac{(-r_k + r_{k+N/2} + x_k - a_1 x_k)u_k + (r_k + a_1 r_{k+N/2})x_k}{\sigma^2} \right\}. \quad (12.62)$$

Inserting (12.62) into (12.56), in the case of knowledge of the IPDP we get for the log-likelihood ratio

$$
\begin{aligned}
L = \sum_{k=1}^{N/2} & \left(\ln \left[4 \cosh \left(\frac{(r_k + r_{k+N/2}) x_k}{\sigma^2} \right) \cosh \left(\frac{(r_k - r_{k+N/2}) u_k}{\sigma^2} \right) + 4 \cosh \left(\frac{2 x_k u_k}{\sigma^2} \right) \right. \right. \\
& \left. \left. \cdot \left(\cosh \left(\frac{(r_k + r_{k+N/2})(u_k + x_k)}{\sigma^2} \right) + \cosh \left(\frac{(r_k + r_{k+N/2})(u_k - x_k)}{\sigma^2} \right) \right) \right] \right) \\
- \sum_{k=1}^{N/2} & \left(\ln \left[4 \cosh \left(\frac{(r_k + r_{k+N/2}) u_k}{\sigma^2} \right) \cosh \left(\frac{(r_k - r_{k+N/2}) x_k}{\sigma^2} \right) + 4 \cosh \left(\frac{2 x_k u_k}{\sigma^2} \right) \right. \right. \\
& \left. \left. \cdot \left(\cosh \left(\frac{(r_k - r_{k+N/2})(u_k + x_k)}{\sigma^2} \right) + \cosh \left(\frac{(r_k - r_{k+N/2})(u_k - x_k)}{\sigma^2} \right) \right) \right] \right) \quad (12.63)
\end{aligned}
$$

Without ISI, i.e., $\vec{u} = 0$, the log-likelihood ratio (12.63) reduces to

$$
L = \sum_{k=1}^{N/2} \left(\ln \left(\cosh(\iota_k) \right) - \ln \left(\cosh(\kappa_k) \right) \right) \quad (12.64)
$$

with

$$
\iota_k = \frac{(r_k + r_{k+N/2}) x_k}{\sigma^2} \quad (12.65)
$$

and

$$
\kappa_k = \frac{(r_k - r_{k+N/2}) x_k}{\sigma^2}. \quad (12.66)
$$

The decision rule can be explained easily for the noiseless case as follows: If $a_1 = 1$ then $\kappa_k = 0$ while ι_k is non-zero. Since $\cosh(0) = 1$ and since $\cosh(x) \geq 1$ the second term in (12.64) becomes 0. Thus, $L > 0$ if $a_1 = 1$. Analog for $a_1 = -1$ it follows that $\iota_k = 0$ while κ_k is non-zero and thus, the first term in (12.64) becomes 0. In such a case it can be observed that $L < 0$. The receiver with consideration of the ISI is denoted in the following as $\text{ML}_{\text{TR,IPDP,ISI}}$ while the latter one is referred to as $\text{ML}_{\text{TR,IPDP}}$.

Average Power Delay Profile

As already assumed in the previous section the receiver knows the correlation matrices of $\vec{h}$ and $\vec{g}$, which are given by

$$\Lambda_{hh} = \begin{bmatrix} \mathcal{E}[h_1^2] & \cdots & 0 \\ \vdots & \ddots & \vdots \\ 0 & \cdots & \mathcal{E}[h_{N/2}^2] \end{bmatrix} = \begin{bmatrix} \lambda_{h,1} & \cdots & 0 \\ \vdots & \ddots & \vdots \\ 0 & \cdots & \lambda_{h,N/2} \end{bmatrix} \tag{12.67}$$

and

$$\Lambda_{gg} = \begin{bmatrix} \mathcal{E}[g_1^2] & \cdots & 0 \\ \vdots & \ddots & \vdots \\ 0 & \cdots & \mathcal{E}[g_{N/2}^2] \end{bmatrix} = \begin{bmatrix} \lambda_{g,1} & \cdots & 0 \\ \vdots & \ddots & \vdots \\ 0 & \cdots & \lambda_{g,N/2} \end{bmatrix}, \tag{12.68}$$

respectively. The receive vector is again rearranged such that the kth tap of $\vec{r}_1$ and $\vec{r}_2$ are adjacent, i.e.,

$$\vec{r} = \begin{bmatrix} r_1, r_{N/2+1}, r_2, r_{N/2+2}, \ldots, r_{N/2-1}, r_N \end{bmatrix}. \tag{12.69}$$

Hence, the correlation matrix Λ_{rr} can be written as

$$\Lambda_{rr} = \begin{bmatrix} \lambda_{r,11} & \lambda_{r,12} & \cdots & 0 & 0 \\ \lambda_{r,21} & \lambda_{r,22} & \cdots & 0 & 0 \\ \vdots & \vdots & \ddots & \vdots & \vdots \\ 0 & 0 & \cdots & \lambda_{r,N/2-1N/2-1} & \lambda_{r,N/2-1N/2} \\ 0 & 0 & \cdots & \lambda_{r,N/2N/2-1} & \lambda_{r,N/2N/2} \end{bmatrix} \tag{12.70}$$

with

$$\lambda_{r,kk} = \lambda_{hk} + \lambda_{gk} + \sigma^2 \tag{12.71}$$

for odd k's and

$$\lambda_{r,kk} = \lambda_{hk-1} + \lambda_{gk-1} + \sigma^2 \tag{12.72}$$

for even k's. The correlated elements on the secondary diagonals are given by

$$\lambda_{r,kk+1} = \lambda_{r,k+1k} = a_1\lambda_{hk} + a_0\lambda_{gk}. \tag{12.73}$$

According to [111], the desired probability assuming the knowledge of the correlation matrix Λ_{rr} is defined as

$$p\left(\vec{r}|a_1 = a_i, a_0, \Lambda_{rr}\right) = \left(\frac{1}{\sqrt{2\pi}}\right)^N \frac{1}{\sqrt{\Delta_{rr}}} \exp\left(-\frac{1}{2}\vec{r}^T\Lambda_{rr}^{-1}\vec{r}\right) \tag{12.74}$$

where Δ_{rr} denotes the determinant of Λ_{rr} and $\vec{r}^T$ is the transpose of $\vec{r}$. The determinant Δ_{rr} can be obtained as

$$\Delta_{rr} = \sum_{k=1}^{N/2}\left[\left(\lambda_{gk} + \lambda_{hk} + \sigma^2\right)^2 - \left(\lambda_{gk}a_0 + \lambda_{hk}a_1\right)^2\right]. \tag{12.75}$$

The inverse of Λ_{rr} has the same structure as (12.70) with the elements

$$(\Lambda_{rr})_{kk}^{-1} = \frac{1}{2(\lambda_{gk} + \lambda_{hk} - \lambda_{gk}a_0 - \lambda_{hk}a_1 + \sigma^2)} + \frac{1}{2(\lambda_{gk} + \lambda_{hk} + \lambda_{gk}a_0 + \lambda_{hk}a_1 + \sigma^2)} \tag{12.76}$$

for odd k's and

$$(\Lambda_{rr})_{kk}^{-1} = \frac{1}{2(\lambda_{gk-1} + \lambda_{hk-1} - \lambda_{gk-1}a_0 - \lambda_{hk-1}a_1 + \sigma^2)} + \frac{1}{2(\lambda_{gk-1} + \lambda_{hk-1} + \lambda_{gk-1}a_0 + \lambda_{hk-1}a_1 + \sigma^2)} \tag{12.77}$$

for even k's. The elements on the secondary diagonal of the inverse $(\Lambda_{rr})^{-1}$ can be derived as

$$(\Lambda_{rr})_{kk+1}^{-1} = \frac{1}{2(\lambda_{gk} + \lambda_{hk} - \lambda_{gk}a_0 - \lambda_{hk}a_1 + \sigma^2)} + \frac{1}{2(\lambda_{gk} + \lambda_{hk} + \lambda_{gk}a_0 + \lambda_{hk}a_1 + \sigma^2)}. \tag{12.78}$$

Evaluating (12.74) and averaging over a_0, the log-likelihood ratio for ML$_{\text{TR,APDP,ISI}}$ can be derived as

$$
\begin{aligned}
L = \ln &\left(\frac{\exp\left\{ -\frac{1}{4} \sum_{k=1}^{N/2} \left[\frac{(r_{2k-1}+r_{2k})^2}{\sigma^2+2\lambda_{gk}+2\lambda_{hk}} + \frac{(r_{2k-1}-r_{2k})^2}{\sigma^2} \right] \right\}}{\sqrt{\sum_{k=1}^{N/2} \sigma^4 + 2\lambda_{gk}\sigma^2 + 2\lambda_{hk}\sigma^2}} \right. \\
&+ \left. \frac{\exp\left\{ -\frac{1}{4} \sum_{k=1}^{N/2} \left[\frac{(r_{2k-1}+r_{2k})^2}{\sigma^2+2\lambda_{hk}} + \frac{(r_{2k-1}-r_{2k})^2}{\sigma^2+2\lambda_{gk}} \right] \right\}}{\sqrt{\sum_{k=1}^{N/2} \sigma^4 + 2\lambda_{gk}\sigma^2 + 2\lambda_{hk}\sigma^2 + 4\lambda_{hk}\lambda_{gk}}} \right) \\
-\ln &\left(\frac{\exp\left\{ -\frac{1}{4} \sum_{k=1}^{N/2} \left[\frac{(r_{2k-1}+r_{2k})^2}{\sigma^2+2\lambda_{gk}} + \frac{(r_{2k-1}-r_{2k})^2}{\sigma^2+2\lambda_{hk}} \right] \right\}}{\sqrt{\sum_{k=1}^{N/2} \sigma^4 + 2\lambda_{gk}\sigma^2 + 2\lambda_{hk}\sigma^2 + 4\lambda_{hk}\lambda_{gk}}} \right. \\
&+ \left. \frac{\exp\left\{ -\frac{1}{4} \sum_{k=1}^{N/2} \left[\frac{(r_{2k-1}+r_{2k})^2}{\sigma^2} + \frac{(r_{2k-1}-r_{2k})^2}{\sigma^2+2\lambda_{gk}+2\lambda_{hk}} \right] \right\}}{\sqrt{\sum_{k=1}^{N/2} \sigma^4 + 2\lambda_{gk}\sigma^2 + 2\lambda_{hk}\sigma^2}} \right) .
\end{aligned}
\tag{12.79}
$$

In the case, where no ISI is present, i.e., $\lambda_{gk} = 0$, we get the ML$_{\text{TR,APDP}}$ while the log-likelihood ratio is given by

$$
L = \frac{1}{\sigma^2} \sum_{k=1}^{N/2} \frac{r_k r_{k+N/2}}{2 + \frac{\sigma^2}{\lambda_k}}
\tag{12.80}
$$

This decision rule is the same as in (12.50) extended by a weighting with the noise variance and the average energy of the kth channel tap. If the noise variance σ_n^2 is large compared to λ_k^2, the multiplication of reference and data signal for the kth channel tap does not contribute to L.

No Channel State Information

Finally, the case is considered again in which the receiver only knows the average energy of the CIR and the noise variance σ^2. As described in section 12.2.2, the elements on the diagonals of the correlation matrices (12.67) and (12.68) are assumed to be equal, i.e., $\forall k : \lambda_{h,k} = \lambda_{g,k} = \lambda'$.

Then, (12.79) results in

$$
\begin{aligned}
L = \ln & \left(\frac{\exp\left\{ -\frac{1}{4} \sum_{k=1}^{N/2} \left[\frac{(r_{2k-1}+r_{2k})^2}{\sigma^2+4\lambda'} + \frac{(r_{2k-1}-r_{2k})^2}{\sigma^2} \right] \right\}}{\sqrt{\sum_{k=1}^{N/2} \sigma^4 + 4\lambda'\sigma^2}} \right. \\
& + \left. \frac{\exp\left\{ -\frac{1}{4} \sum_{k=1}^{N/2} \left[\frac{2(r_{2k-1}^2+r_{2k}^2)}{\sigma^2+2\lambda'} \right] \right\}}{\sqrt{\sum_{k=1}^{N/2} (\sigma^2+2\lambda')^2}} \right) \\
& - \ln \left(\frac{\exp\left\{ -\frac{1}{4} \sum_{k=1}^{N/2} \left[\frac{2(r_{2k-1}^2+r_{2k}^2)}{\sigma^2+2\lambda'} \right] \right\}}{\sqrt{\sum_{k=1}^{N/2} (\sigma^2+2\lambda')^2}} \right. \\
& + \left. \frac{\exp\left\{ -\frac{1}{4} \sum_{k=1}^{N/2} \left[\frac{(r_{2k-1}+r_{2k})^2}{\sigma^2} + \frac{(r_{2k-1}-r_{2k})^2}{\sigma^2+4\lambda'} \right] \right\}}{\sqrt{\sum_{k=1}^{N/2} \sigma^4 + 4\lambda'\sigma^2}} \right) .
\end{aligned}
\tag{12.81}
$$

In the case where no ISI is present, the ML receiver from (12.80) reduces to the typical transmitted reference receiver, i.e,

$$
\begin{aligned}
L &= \frac{1}{\sigma^2(2+\frac{\sigma^2}{\lambda'})} \sum_{k=1}^{N/2} r_k r_{k+N/2} \\
&\propto \sum_{k=1}^{N/2} r_k r_{k+N/2},
\end{aligned}
\tag{12.82}
$$

which is introduced in section 4.7.2. This receiver is referred to as TR in the following.

12.3.3 Performance Results

As in section 12.2.3, the different receiver structures are compared by means of their BER curves. The channel taps are also assumed to be uniformly distributed within the duration of the channel impulse response. Furthermore, one transmitted reference doublet has to be transmitted in a frame of $T = 20$ ns to achieve a data rate of 50 Mbps. We consider again the cases that no ISI, weak ISI, and strong ISI occur with CIR durations of 10 ns, 14 ns, and 17 ns, respectively.

No ISI In Fig. 12.6, the BER curves are plotted for the different ML receivers assuming no ISI. Again, the performance increases with higher level of CSI. The $\text{ML}_{\text{TR,Full}}$ performs a coherent detection, where the data pulse is correlated with a template. In the case of full CSI, the reference pulse is not required for the detection. The $ML_{TR,IPDP}$ is about 4 dB worse compared to the $ML_{TR,Full}$. This performance degradation is caused due to the missing knowledge of the channel tap signs which makes a coherent combining impossible. The $ML_{TR,APDP}$ equals the TR receiver in the case of uniformly distributed channel taps. Therefore, both receiver structures have the same performance. However, it is interesting to see that the knowledge of the IPDP results in a gain of only about 1.5 dB compared to the TR receiver.

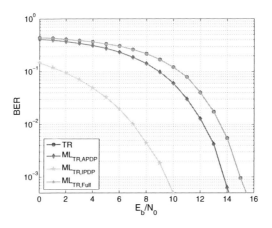

Figure 12.6: Bit error rate curves in the case of no ISI

Weak ISI   In the presence of weak ISI the ML receiver structures not considering the ISI degrade as it can be seen in Fig. 12.7. While the degradation of the $ML_{TR,Full}$ and $ML_{TR,IPDP}$ is only minor, the degradation of the $ML_{TR,APDP}$ and of the TR is severe. However, the performance degradation of the $ML_{TR,APDP}$ can be reduced by considering the ISI. The $ML_{TR,APDP,ISI}$ is several dB better than the $ML_{TR,APDP}$. As in the no ISI case the TR receiver and the $ML_{TR,APDP}$ exhibit the same performance.

Strong ISI   The $ML_{TR,Full}$ yields a good performance even in the presence of strong ISI as it is shown in Fig. 12.8. This receiver is only slightly worse than the $ML_{TR,Full,ISI}$. In the presence of

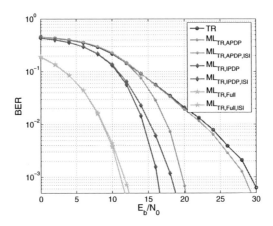

Figure 12.7: Bit error rate curves in the case of weak ISI

strong ISI the $ML_{TR,IPDP,ISI}$ performs much better than the $ML_{TR,IPDP}$. The $ML_{TR,APDP,ISI}$ yields substantially better results than the normal $ML_{TR,APDP}$. This receiver and the TR receiver both approach an error floor at $6 \cdot 10^{-2}$, which makes them not suitable for the strong ISI scenario. Hence, the simple TR receiver is only suited for applications with moderate data rates, where adjacent pulses do not cause too much ISI. For applications requiring high data rates or in the case of channels with a large delay spread, TR receivers with partial CSI are preferable. However, this performance gain is achieved on cost of complexity.

Conclusions   Similar to the derivations in section 12.2 a family of symbol-wise detectors for a transmitted reference scheme was presented, which utilizes partial channel state information to improve the robustness against ISI. Again, the cases of no CSI, APDP, IPDP, and full CSI were considered. As for the binary PPM ML receivers the performance of the TR ML receivers decreased in the presence of ISI. However, a performance gain can be achieved by considering the ISI for the derivation of the ML receivers. Such receivers exhibit a good performance even in the presence of strong ISI. This performance gain is achieved at cost of a higher receiver complexity and a necessary channel estimation. Therefore, the usability of such receivers for

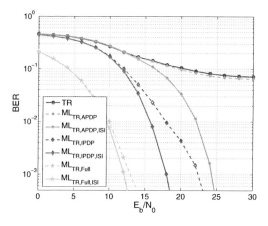

Figure 12.8: Bit error rate curves in the case of strong ISI

low complexity devices is very limited. The TR receiver, which is the most promising TR receiver structure for the use in very simple WBANs, suffers from strong ISI and can only be used in the presence of moderate ISI. However, this receiver has a higher complexity compared to the ED, which makes the latter one more preferable.

12.4 Maximum Likelihood Receivers for Transmitted Reference Systems in the Presence of CCI

In the previous sections, the maximum likelihood receivers for binary PPM and transmitted reference systems were derived for different levels of CSI. In the following, a transmission without ISI is assumed. However, the impact of a co-channel interferer is evaluated and the maximum likelihood receivers with different CSI level are derived for the transmitted reference system, exemplarily.

12.4.1 System Description

There are two TR systems considered, each of the being transmitted one reference and one data pulse per bit. One system is the desired user while the other system is considered as a CCI. Both systems are assumed to be synchronous. The two signals at the receiver are given by

$$\vec{r}_1 = \vec{h}_U + \vec{h}_I + \vec{n}_1 \tag{12.83}$$

for the reference pulses and

$$\vec{r}_2 = \vec{h}_U \cdot a_U + \vec{h}_I \cdot a_I + \vec{n}_2 \tag{12.84}$$

for the data pulses. $\vec{h}_U$ and $\vec{h}_I$ denote the channel of the desired and the interference signal, respectively, and $\vec{n}_2$ and $\vec{n}_1$ contain the additive white Gaussian noise (AWGN) both with variance σ_D^2 and σ_R^2, respectively. The transmitted bits a_U and a_I can each be either +1 or -1. To determine the ML estimator the probability $p(\vec{r}_1, \vec{r}_2 | a_U = a_i, C_U, C_I)$ has to be maximized. C_U denotes the level of CSI for the desired user that is known at the receiver side and C_I denotes the level of CSI for the interfering user. In the following it is assumed that the receiver has the same amount of CSI for the desired and for the interfering user.

12.4.2 ML Estimation with full CSI of desired signal and CCI

In a first step, perfect CSI is assumed at the receiver side, i.e., $C_U = \vec{h}_U$ and $C_I = \vec{h}_I$. Thus, the probability that has to be maximized is given by

$$p(\vec{r}_1, \vec{r}_2 | a_U = a_i, a_I, \vec{h}_U, \vec{h}_I)$$
$$= p(\vec{n}_1 = \vec{r}_1 - \vec{h}_U - \vec{h}_I, \vec{n}_2 - \vec{h}_U a_U - \vec{h}_I a_I) \tag{12.85}$$

with the PDFs for the noise terms defined as

$$p(n_{1,k}) = \frac{1}{\sqrt{2\pi\sigma_1^2}} \cdot \exp\left\{ -\frac{n_{1,k}^2}{2\sigma_1^2} \right\} \tag{12.86}$$

and

$$p(n_{2,k}) = \frac{1}{\sqrt{2\pi\sigma_2^2}} \cdot \exp\left\{ -\frac{n_{2,k}^2}{2\sigma_2^2} \right\} \tag{12.87}$$

Due to the independence of n_1 and n_2 and by assuming that $\sigma_1^2 = c \cdot \sigma_2^2$, (12.85) yields

$$
\begin{aligned}
& p(\vec{r}_1, \vec{r}_2 | a_U = a_i, \vec{h}_U, \vec{h}_I, a_I) \\
& = \left(\frac{1}{2\pi\sigma_1\sigma_2} \right)^N \cdot \exp \left\{ -\sum_{k=1}^{N} \frac{(r_{1,k} - h_{U,k} - h_{I,k})^2}{2c \cdot \sigma_2^2} + \frac{(r_{2,k} - h_{U,k}a_U - h_{I,k}a_I)^2}{2\sigma_2^2} \right\},
\end{aligned}
$$

(12.88)

where c can be arbitrary but fixed. Since the transmit bit a_I of the interferer is not known at the receiver, it has to be averaged out. With $a_I \in \{-1; +1\}$, the distribution of a_I is given by

$$
p(a_I) = \frac{1}{2}\delta(a_I - 1) + \frac{1}{2}\delta(a_I + 1).
$$

(12.89)

Considering only the terms in (12.88) relevant for the log-likelihood ratio one gets

$$
\begin{aligned}
& p(\vec{r}_1, \vec{r}_2 | a_U = a_i, \vec{h}_U, \vec{h}_I) \\
& \propto \int_{-\infty}^{\infty} p(a_I) \cdot \exp \left\{ \sum_{k=1}^{N} \frac{h_{U,k}a_U r_{2,k} + h_{I,k}a_I r_{2,k} - h_{U,k}h_{I,k}a_U a_I}{\sigma_2^2} \right\} da_I \\
& = \frac{1}{2} \exp \left\{ \sum_{k=1}^{N} \frac{h_{U,k}a_U r_{2,k}}{\sigma_2^2} \right\} \\
& \quad \cdot \left(\exp \left\{ \sum_{k=1}^{N} \frac{h_{I,k}r_{2,k} - h_{U,k}h_{I,k}a_U}{\sigma_2^2} \right\} + \exp \left\{ -\sum_{k=1}^{N} \frac{h_{I,k}r_{2,k} - h_{U,k}h_{I,k}a_U}{\sigma_2^2} \right\} \right)
\end{aligned}
$$

(12.90)

Due to $\frac{\exp\{x\}+\exp\{-x\}}{2} = \cosh\{x\}$, (12.90) can be written as

$$
p(\vec{r}_1, \vec{r}_2 | a_U = a_i, \vec{h}_U, \vec{h}_I) = \exp \left\{ \sum_{k=1}^{N} \frac{h_{U,k}a_U r_{2,k}}{\sigma_2^2} \right\} \cdot \prod_{k=1}^{N} \cosh \left\{ \frac{h_{I,k}r_{2,k} - h_{U,k}h_{I,k}a_U}{\sigma_2^2} \right\}.
$$

(12.91)

The likelihood ratio M with full CSI is given by

$$
M = \frac{p(\vec{r}_1, \vec{r}_2 | a_U = 1, \vec{h}_U, \vec{h}_I) \cdot P[s = 1]}{p(\vec{r}_1, \vec{r}_2 | a_U = -1, \vec{h}_U, \vec{h}_I) \cdot P[s = -1]}
$$

(12.92)

and the log-likelihood ratio L by

$$
\begin{aligned}
L & = \ln(M) \\
& = \ln \left(\frac{P[a_U = 1]}{P[a_U = -1]} \right) + \ln \left(p(\vec{r}_1, \vec{r}_2 | a_U = 1, \vec{h}) \right) - \ln \left(p(\vec{r}_1, \vec{r}_2 | a_U = 1, \vec{h}) \right).
\end{aligned}
$$

(12.93)

With $P[a_U = 1] = P[a_U = -1]$, the log-likelihood becomes

$$
\begin{aligned}
L = & 2 \sum_{k=1}^{N} \frac{h_{U,k} r_{2,k}}{\sigma_2^2} \\
& + \sum_{k=1}^{N} \left[\ln \left(\cosh \left\{ \frac{h_{I,k} r_{2,k} - h_{U,k} h_{I,k}}{\sigma_2^2} \right\} \right) - \ln \left(\cosh \left\{ \frac{h_{I,k} r_{2,k} + h_{U,k} h_{I,k}}{\sigma_2^2} \right\} \right) \right].
\end{aligned}
$$

$$(12.94)$$

Compared to (12.58), where no CCI is present, the influence of the interferer is subtracted from the correlation of the data signal with the desired channel h_1.

12.4.3 ML estimation with knowledge of the IPDP of the desired signal and the CCI

Full CSI is usually not available at the receiver side. Hence, we derive the maximum likelihood estimator assuming the knowledge of the IPDP of desired signal and CCI. The channel $\vec{h}_i$ can be described by $\vec{h}_i = \vec{x}_i \odot \vec{z}_i$ with $z_{i,k} \in \{-1; 1\}$ as the sign of the kth channel tap $x_{i,k}$ and $\odot$ denoting the element-wise multiplication. This yields $\mathcal{C}_U = \left| \vec{h}_U \right| = \vec{x}_U$ and $\mathcal{C}_I = \left| \vec{h}_I \right| = \vec{x}_I$. Thus, the probability that has to be maximized is given by

$$
p(\vec{r}_1, \vec{r}_2 | a_U = a_i, \vec{x}_U, \vec{x}_I). \tag{12.95}
$$

Since the signs $\vec{z}_U$ and $\vec{z}_I$ of the channel taps are not known, they have to be averaged out yielding

$$
\begin{aligned}
& p(\vec{r}_1, \vec{r}_2 | a_U = a_i, \vec{x}_U, \vec{x}_I) \\
& = \int_{-\infty}^{\infty} p(a_I) \cdot \int_{-\infty}^{\infty} \prod_{k=1}^{N} p(z_{2,k}) \cdot \int_{-\infty}^{\infty} \prod_{k=1}^{N} p(z_{1,k}) \cdot p(\vec{r}_1, \vec{r}_2 | a_U = a_i, \vec{h}_U, \vec{h}_I) dz_U dz_I da_I.
\end{aligned}
$$

$$(12.96)$$

With (12.89), $\sigma_1^2 = c \cdot \sigma_2^2$, and

$$
P[z_{i,k}] = \frac{1}{2} \delta(z_{i,k} - 1) + \delta(z_{i,k} + 1) \tag{12.97}
$$

(12.96) can be written as

$$
p(\vec{r}_1, \vec{r}_2 | a_U = a_i, \vec{x}_U, \vec{x}_I) \propto \prod_{k=1}^{N} \exp\left\{ -\frac{r_{2,k}(x_{I,k} + a_U x_{U,k}(1 + x_{I,k}))}{2\sigma_2^2} \right\}
$$

$$
\cdot \left[1 + \exp\left\{ \frac{r_{1,k}(1 + x_{U,k}) x_{I,k}}{c\sigma_2^2} \right\} + \exp\left\{ \frac{r_{2,k}(1 + a_U x_{U,k}) x_{I,k}}{\sigma_2^2} \right\} \right.
$$

$$
+ \exp\left\{ \frac{(r_{1,k}(1 + x_{U,k})\sigma_2^2 + r_{2,k} c\sigma_2^2 (1 + a_U x_{U,k})) x_{I,k}}{c\sigma_2^4} \right\}
$$

$$
+ \exp\left\{ \frac{(r_{1,k}\sigma_2^2 + r_{2,k} a_U c\sigma_2^2) x_{U,k}(1 + x_{I,k})}{c\sigma_2^4} \right\}
$$

$$
+ \exp\left\{ \frac{r_{1,k}(x_{U,k} + x_{I,k}) + cr_{2,k} a_U x_{U,k}(1 + x_{I,k})}{c\sigma_2^2} \right\}
$$

$$
+ \exp\left\{ \frac{r_{1,k} x_{U,k}(1 + x_{I,k}) + cr_{2,k}(a_U x_{U,k} + x_{I,k})}{c\sigma_2^2} \right\}
$$

$$
\left. + \exp\left\{ \frac{r_{1,k}(x_{U,k} + x_{I,k}) + cr_{2,k}(a_U x_{U,k} + x_{I,k})}{c\sigma_2^2} \right\} \right] \tag{12.98}
$$

considering only terms relevant for the log-likelihood ratio. Assuming $P[s = 1] = P[s = -1]$, the log-likelihood becomes

$$
L = P_{a=1} - P_{a=-1} \tag{12.99}
$$

with

$$
P_{a=1} = \sum_{k=1}^{N} \ln\left(\exp\left\{ -\frac{r_{2,k}(x_{I,k} + x_{U,k}(1 + x_{I,k}))}{2\sigma_2^2} \right\} \right.
$$

$$
\cdot \left[1 + \exp\left\{ \frac{r_{1,k}(1 + x_{U,k}) x_{I,k}}{c\sigma_2^2} \right\} + \exp\left\{ \frac{r_{2,k}(1 + x_{U,k}) x_{I,k}}{\sigma_2^2} \right\} \right.
$$

$$
+ \exp\left\{ \frac{(r_{1,k}(1 + x_{U,k})\sigma_2^2 + r_{2,k} c\sigma_2^2 (1 + x_{U,k})) x_{I,k}}{c\sigma_2^4} \right\}
$$

$$
+ \exp\left\{ \frac{(r_{1,k}\sigma_2^2 + r_{2,k} c\sigma_2^2) x_{U,k}(1 + x_{I,k})}{c\sigma_2^4} \right\}
$$

$$
+ \exp\left\{ \frac{r_{1,k}(x_{U,k} + x_{I,k}) + cr_{2,k} x_{U,k}(1 + x_{I,k})}{c\sigma_2^2} \right\}
$$

$$
+ \exp\left\{ \frac{r_{1,k} x_{U,k}(1 + x_{I,k}) + cr_{2,k}(x_{U,k} + x_{I,k})}{c\sigma_2^2} \right\}
$$

$$
\left. \left. + \exp\left\{ \frac{r_{1,k}(x_{U,k} + x_{I,k}) + cr_{2,k}(x_{U,k} + x_{I,k})}{c\sigma_2^2} \right\} \right] \right) \tag{12.100}
$$

and

$$P_{a=-1} = \sum_{k=1}^{N} \ln \left(\exp \left\{ -\frac{r_{2,k}(x_{I,k} - x_{U,k}(1 + x_{I,k}))}{2\sigma_2^2} \right\} \right.$$

$$\cdot \left[1 + \exp \left\{ \frac{r_{1,k}(1 + x_{U,k})x_{I,k}}{c\sigma_2^2} \right\} + \exp \left\{ \frac{r_{2,k}(1 - x_{U,k})x_{I,k}}{\sigma_2^2} \right\} \right.$$

$$+ \exp \left\{ \frac{(r_{1,k}(1 + x_{U,k})\sigma_2^2 + r_{2,k}c\sigma_2^2(1 - x_{U,k}))x_{I,k}}{c\sigma_2^4} \right\}$$

$$+ \exp \left\{ \frac{(r_{1,k}\sigma_2^2 - r_{2,k}c\sigma_2^2)x_{U,k}(1 + x_{I,k})}{c\sigma_2^4} \right\}$$

$$+ \exp \left\{ \frac{r_{1,k}(x_{U,k} + x_{I,k}) - cr_{2,k}x_{U,k}(1 + x_{I,k})}{c\sigma_2^2} \right\}$$

$$+ \exp \left\{ \frac{r_{1,k}x_{U,k}(1 + x_{I,k}) + cr_{2,k}(-x_{U,k} + x_{I,k})}{c\sigma_2^2} \right\}$$

$$\left. \left. + \exp \left\{ \frac{r_{1,k}(x_{U,k} + x_{I,k}) + cr_{2,k}(-x_{U,k} + x_{I,k})}{c\sigma_2^2} \right\} \right] \right). \tag{12.101}$$

Without CCI the log-likelihood ratio from (12.99) simplifies substantially resulting in the log-likelihood rule given in (12.64).

12.4.4 Maximum likelihood estimation with knowledge of the APDP of the desired signal and the CCI

In this section the CSI is limited to the knowledge of the APDP, i.e., $C_U = \Lambda_{hUhU}$ and $C_I = \Lambda_{hIhI}$, where Λ_{hihi} denotes the correlation matrix of the channel h_i and is given as diagonal matrix

$$\Lambda_{hihi} = \begin{bmatrix} \mathcal{E}\left[h_{i,1}^2\right] & \cdots & 0 \\ \vdots & \ddots & \vdots \\ 0 & \cdots & \mathcal{E}\left[h_{i,N}^2\right] \end{bmatrix} = \begin{bmatrix} \lambda_{i,1} & \cdots & 0 \\ \vdots & \ddots & \vdots \\ 0 & \cdots & \lambda_{i,N} \end{bmatrix}. \tag{12.102}$$

As described in (12.69), the receive signals are reordered such that the correlated elements of d_R and d_D are adjacent, i.e.,

$$\vec{r} = [\vec{r}_1[1], \vec{r}_2[1], \vec{r}_1[2], \vec{r}_2[2], \ldots, \vec{r}_1[N], \vec{r}_2[N]]$$

$$= [r[1], r[2], \ldots, r[2N-1], r[2N]]. \tag{12.103}$$

With

$$\mu_{1,k} = \lambda_{U,k} + \lambda_{I,k} + \sigma_1^2$$

$$\mu_{2,k} = \lambda_{U,k} + \lambda_{I,k} + \sigma_2^2$$

$$\nu_k = a_U \cdot \lambda_{U,k} + a_I \cdot \lambda_{I,k} \tag{12.104}$$

the correlation matrix Λ_{rr} of $\vec{r}$ is given by

$$\Lambda_{rr} = \mathcal{E}_{h,n|a_U,a_I}\left[\vec{r}\vec{r}^T\right] = \begin{bmatrix} \mu_{1,1} & \nu_1 & 0 & \dots & 0 \\ \nu_1 & \mu_{2,1} & 0 & \ddots & \vdots \\ 0 & 0 & \ddots & 0 & 0 \\ \vdots & \ddots & 0 & \mu_{1,N} & \nu_N \\ 0 & \dots & 0 & \nu_N & \mu_{2,N} \end{bmatrix}. \tag{12.105}$$

According to [111] the probability of $\vec{r}$ given the correlation matrix can be derived as

$$p\left(\vec{r}|a_U = a_{Ui}, a_I = a_{Ii}, \Lambda_{rr}\right)$$
$$= \left(\frac{1}{\sqrt{2\pi}}\right)^N \cdot \frac{1}{\sqrt{\Delta_{rr}}} \cdot \exp\left(-\frac{1}{2}\vec{r}^T \Lambda_{rr}^{-1} \vec{r}\right), \tag{12.106}$$

with Δ_{rr} denoting the determinant of Λ_{rr}. Since a_I is not known at the receiver, it has to be averaged out and (12.106) becomes

$$p\left(\vec{r}|a_U = a_{Ui}, \Lambda_{rr}\right)$$
$$= \frac{1}{2} \cdot \left(\frac{1}{\sqrt{2\pi}}\right)^N \cdot \frac{1}{\sqrt{\Delta_{rr,a_{Ii}=1}}} \cdot \exp\left(-\frac{1}{2}\vec{r}^T \Lambda_{rr}^{-1} \vec{r}\right)_{a_{Ii}=1}$$
$$+ \frac{1}{2} \cdot \left(\frac{1}{\sqrt{2\pi}}\right)^N \cdot \frac{1}{\sqrt{\Delta_{rr,a_{Ii}=-1}}} \cdot \exp\left(-\frac{1}{2}\vec{r}^T \Lambda_{rr}^{-1} \vec{r}\right)_{a_{Ii}=-1}. \tag{12.107}$$

Assuming $P[s = 1] = P[s = -1]$ and $\sigma_1^2 = c \cdot \sigma_2^2$, this yields for the log-likelihood ratio

$$
L = \ln \left[\frac{\exp\left\{ -\sum_{k=1}^{N} \frac{(r_{2k-1}-r_{2k})^2 \eta_k + (r_{2k-1}^2 + r_{2k}^2 c)\sigma_2^2}{\xi_k \sigma_2^2} \right\}}{2\sqrt{\prod_{k=1}^{N} \xi_k \sigma_2^2}} \right.
$$
$$
\left. + \frac{\exp\left\{ \sum_{k=1}^{N} \frac{2r_{2k-1}r_{2k}\kappa_k - r_{2k-1}^2(\eta_k+\sigma_2^2) - r_{2k}^2(\eta_k+c\sigma_2^2)}{\zeta_k} \right\}}{2\sqrt{\prod_{k=1}^{N} \zeta_k}} \right]
$$
$$
- \ln \left[\frac{\exp\left\{ -\sum_{k=1}^{N} \frac{(r_{2k-1}+r_{2k})^2 \eta_k + (r_{2k-1}^2 + r_{2k}^2 c)\sigma_2^2}{\xi_k \sigma_2^2} \right\}}{2\sqrt{\prod_{k=1}^{N} \xi_k \sigma_2^2}} \right.
$$
$$
\left. + \frac{\exp\left\{ \sum_{k=1}^{N} \frac{-2r_{2k-1}r_{2k}\kappa_k - r_{2k-1}^2(\eta_k+\sigma_2^2) - r_{2k}^2(\eta_k+c\sigma_2^2)}{\zeta_k} \right\}}{2\sqrt{\prod_{k=1}^{N} \zeta_k}} \right]
$$

$$(12.108)$$

with

$$\eta_k = \lambda_{U,k} + \lambda_{I,k},$$
$$\kappa_k = \lambda_{U,k} - \lambda_{I,k},$$
$$\xi_k = (1 + c)\eta_k + c\sigma_2^2,$$
$$\zeta_k = 4\lambda_{U,k}\lambda_{I,k} + (1 + c)\eta_k\sigma_2^2 + c\sigma_2^4. \tag{12.109}$$

The result for a TR ML estimator without CCI considering the APDP is given in (12.80).

12.4.5 Maximum Likelihood Estimation without CSI

Finally, the ML estimator, which only knows noise variances and the average transmit power, is derived again as a special case of the ML receiver with APDP knowledge. In this case, all elements of the correlation matrix in (12.102) equal λ_U' and λ_I' for the desired user and the interferer, respectively. Thus, (12.108) is given by

$$
L = \ln \left[\frac{\exp\left\{ -\sum_{k=1}^{N} \frac{(r_{2k-1} - r_{2k})^2 \eta' + (r_{2k-1}^2 + r_{2k}^2 c)\sigma_2^2}{\xi' \sigma_2^2} \right\}}{2\sqrt{\prod_{k=1}^{N} \xi' \sigma_2^2}} \right.
$$

$$
\left. + \frac{\exp\left\{ \sum_{k=1}^{N} \frac{2r_{2k-1} r_{2k} \kappa' - r_{2k-1}^2 (\eta' + \sigma_2^2) - r_{2k}^2 (\eta' + c\sigma_2^2)}{\zeta'} \right\}}{2\sqrt{\prod_{k=1}^{N} \zeta'}} \right]
$$

$$
- \ln \left[\frac{\exp\left\{ -\sum_{k=1}^{N} \frac{(r_{2k-1} + r_{2k})^2 \eta' + (r_{2k-1}^2 + r_{2k}^2 c)\sigma_2^2}{\xi' \sigma_2^2} \right\}}{2\sqrt{\prod_{k=1}^{N} \xi' \sigma_2^2}} \right.
$$

$$
\left. + \frac{\exp\left\{ \sum_{k=1}^{N} \frac{-2r_{2k-1} r_{2k} \kappa' - r_{2k-1}^2 (\eta' + \sigma_2^2) - r_{2k}^2 (\eta' + c\sigma_2^2)}{\zeta'} \right\}}{2\sqrt{\prod_{k=1}^{N} \zeta'}} \right]
$$

$$(12.110)$$

with

$$
\eta' = \lambda'_U + \lambda'_I,
$$
$$
\kappa' = \lambda'_U - \lambda'_I,
$$
$$
\xi' = (1 + c)\eta' + c\sigma_2^2,
$$
$$
\zeta' = 4\lambda'_U \lambda'_I + (1 + c)\eta' \sigma_2^2 + c\sigma_2^4. \tag{12.111}
$$

The equation (12.110) simplifies substantially assuming no CCI. Therewith, the log-likelihood ratio is given by the typical transmitted reference receiver in (12.50).

12.4.6 Performance Results

To see the impact of CCI and CSI the performance of the above derived ML estimators is evaluated by means of simulation. We consider a non-sparse channel with 100 uniformly channel taps. The BER curves are shown in Fig. 12.9. For these simulations we assume a signal-to-interference ratio SIR = 10dB. As expected, the ML with full CSI shows the best performance and it is about 3dB better compared to the ML with IPDP. Having no CSI available at the receiver results in a performance loss of about 5dB compared to the ML with full CSI. Since the ML with APDP equals the ML without CSI in the case of a uniformly distributed CIR, the performance of these both receiver structures stays the same.

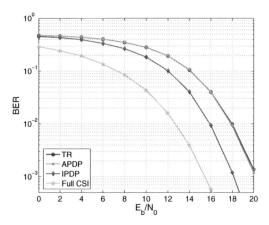

Figure 12.9: Bit error curves for TR ML with different CSI in the presence of a CCI at a SIR of 10dB

Conclusions   The transmitted reference ML receivers in the presence of CCI were derived in the previous sections. As expected, the performance increased with higher level of CSI. However, this performance gain was realized at cost of a required channel estimation and a higher receiver complexity.

12.5 Maximum Likelihood Receivers for TR Pulse Interval Amplitude Modulation

In the previous sections maximum likelihood receivers for binary pulse position modulation and transmitted reference pulse amplitude modulation were derived. In the following both modulation schemes are combined to one modulation scheme, which is referred to as transmitted reference pulse interval amplitude modulation (TR PIAM). In contrast to the typical

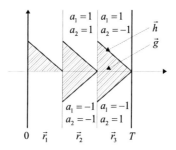

Figure 12.10: Schematic of the considered receive signal positions for TRPIAM

transmitted-reference scheme, where information is contained only in the amplitude, the information is contained in the pulse position, too. It is assumed that the reference pulse is always positive while the sign and the position of the data pulse are determined by the transmit symbol as shown in Fig. 12.10. ISI is omitted in the derivations for the purpose of better clarity. However, it can be easily included analog to the ML derivations for the transmitted reference receiver and energy detector.

As it can be seen from Fig. 12.10 the sampled receive signal in the first time frame is given by

$$\vec{r}_1 = \vec{h} + \vec{n}_1, \tag{12.112}$$

in the second time frame

$$\vec{r}_2 = \frac{1}{2}(a_1 + a_2)\vec{h} + \vec{n}_2, \tag{12.113}$$

and in the third time frame

$$\vec{r}_3 = \frac{1}{2}(a_1 - a_2)\vec{h} + \vec{n}_3. \tag{12.114}$$

The data symbols are $a_0, a_1 \in \{\pm 1\}$ and $\vec{h}$ denotes the channel impulse response. It is assumed that the taps of the CIR are statistically independent normal random variables with zero mean. $\vec{n}_1$, $\vec{n}_2$ and $\vec{n}_3$ contain the additive white Gaussian noise (AWGN), all with variance σ^2. The whole receive signal over the three time slots is given by

$$
\begin{aligned}
\vec{r} &= [\vec{r}_1, \vec{r}_2, \vec{r}_3] \\
&= \vec{h}_1 + \frac{1}{2}(a_1 + a_2)\vec{h}_2 + \frac{1}{2}(a_1 - a_2)\vec{h}_3 + \vec{n}
\end{aligned}
\tag{12.115}
$$

with

$$
\begin{aligned}
\vec{h}_1 &= [\vec{h}, 0, \dots, 0], \\
\vec{h}_2 &= [0, \dots, 0, \vec{h}, 0, \dots, 0], \\
\vec{h}_3 &= [0, \dots, 0, \vec{h}]
\end{aligned}
\tag{12.116}
$$

and

$$
\vec{n} = [\vec{n}_1, \vec{n}_2, \vec{n}_3].
\tag{12.117}
$$

$\vec{r}_1$, $\vec{r}_2$, and $\vec{r}_3$ contain $N/3$ elements each, i.e., $\vec{r}$ contains N elements.

12.5.1 Symbol-wise Maximum Likelihood Detector

The probability $p(\vec{r}|a_1, a_0, \mathcal{C})$ is considered for the derivation of the ML detector as described in section 12.2.2. As noted in the previous sections, the following cases of channel state information $\mathcal{C}$ are considered: (i) full CSI, (ii) IPDP, (iii) APDP, and (iv) no CSI. For the derivation of the maximum likelihood receiver the probability

$$
p\left(\vec{r}|a_1 = a_i, a_2 = a_j, \mathrm{CSI}\right)
\tag{12.118}
$$

has to be calculated as described in section 12.1. Using the TR PIAM scheme with receive signal from (12.115), (12.118) is given by

$$
p\left(\vec{r}|a_1 = a_i, a_2 = a_j, \vec{h}_1, \vec{h}_2\right) = p\left(\vec{n} = \vec{r} - \vec{h}_1 - \frac{1}{2}(a_1 + a_2)\vec{h}_2 - \frac{1}{2}(a_1 - a_2)\vec{h}_3\right).
\tag{12.119}
$$

Assuming statistically independent distributed noise samples with

$$
p(n_k) = \frac{1}{\sqrt{2\pi\sigma^2}} \cdot \exp\left\{\frac{n_k^2}{2\sigma^2}\right\},
\tag{12.120}
$$

(12.119) can be written as

$$
p\left(\vec{r}|a_1 = a_i, a_2 = a_j, \vec{h}_1, \vec{h}_2\right)
$$
$$
= \prod_{k=1}^{N} p\left(n_k = r_k - h_{1,k} - \frac{1}{2}(a_1 + a_2)h_{2,k} - \frac{1}{2}(a_1 - a_2)h_{3,k}\right)
$$
$$
= \prod_{k=1}^{N} \frac{1}{\sqrt{2\pi\sigma^2}} \cdot \exp\left\{-\frac{1}{2\sigma^2}\left(r_k - h_{1,k} - \frac{1}{2}(a_1 + a_2)h_{2,k} - \frac{1}{2}(a_1 - a_2)h_{3,k}\right)^2\right\}.
$$

$$(12.121)$$

Since every time slot has an integer number of samples, the whole number of samples N is a multiple of 3. For the further derivation it is assumed that the channel is invariant of the three adjacent time slots. This means for the channel taps

$$
h_{1,k} = h_{2,k+N/3} = h_{3,k+2N/3} = h_k \tag{12.122}
$$

and allows to rewrite (12.121) as

$$
p\left(\vec{r}|a_1 = a_i, a_2 = a_j, \vec{h}\right) = \prod_{k=1}^{N} \frac{1}{\sqrt{2\pi\sigma^2}} \cdot \exp\left\{-\frac{1}{2\sigma^2}\left(r_k^2 + h_k^2 + \frac{1}{4}(a_1 + a_2)^2 h_k^2\right.\right.
$$
$$
\left.\left. + \frac{1}{4}(a_1 - a_2)^2 h_k^2 - 2r_k h_k - (a_1 + a_2)r_{k+N/3}h_k - (a_1 - a_2)r_{k+2N/3}h_k\right)\right\}
$$
$$
= \prod_{k=1}^{N} \frac{1}{\sqrt{2\pi\sigma^2}} \cdot \exp\left\{-\frac{1}{2\sigma^2}\left(r_k^2 + 2h_k^2 - \left(2r_k + (a_1 + a_2)r_{k+N/3} + (a_1 - a_2)r_{k+2N/3}\right)h_k\right)\right\}.
$$

$$(12.123)$$

Full Channel State Information

As described in the previous sections full CSI at the receiver side is assumed first, i.e., $C_{\text{full}} = [\vec{h}]$. Because of four different hypotheses the likelihood ratio is not suited for TR PIAM. Therefore, the four probabilities for the different hypotheses are calculated in the following. Since the term

$$
\prod_{k=1}^{N} \frac{1}{\sqrt{2\pi\sigma^2}} \cdot \exp\left\{-\frac{1}{2\sigma^2}\left(r_k^2 + 2h_k^2 + 2r_k h_k\right)\right\} \tag{12.124}
$$

is the same for all probabilities, it can be omitted in the following. The probabilities relevant for the determination of the maximum likelihood receiver are given by

$$
p_{1,1} = p\left(\vec{r}|a_1 = 1, a_2 = 1, \vec{h}\right) \propto \exp\left\{\sum_{k=1}^{N/3} \frac{r_{k+N/3}h_k}{\sigma^2}\right\}, \tag{12.125}
$$

$$
p_{1,-1} = p\left(\vec{r}|a_1 = 1, a_2 = -1, \vec{h}\right) \propto \exp\left\{\sum_{k=1}^{N/3} \frac{r_{k+2N/3}h_k}{\sigma^2}\right\}, \tag{12.126}
$$

$$
p_{-1,1} = p\left(\vec{r}|a_1 = -1, a_2 = 1, \vec{h}\right) \propto \exp\left\{\sum_{k=1}^{N/3} \frac{-r_{k+2N/3}h_k}{\sigma^2}\right\}, \tag{12.127}
$$

and

$$
p_{-1,-1} = p\left(\vec{r}|a_1 = -1, a_2 = -1, \vec{h}\right) \propto \exp\left\{\sum_{k=1}^{N/3} \frac{-r_{k+N/3}h_k}{\sigma^2}\right\}. \tag{12.128}
$$

Since the logarithm is strictly monotonic increasing, it is sufficient to consider only the exponents in (12.125) to (12.128) to determine the most likely transmit symbols. The maximum likelihood receiver decides for the symbol with the highest probability out of

$$
\ln\left(p_{1,1}\right) \propto \sum_{k=1}^{N/3} \frac{r_{k+N/3}h_k}{\sigma^2}, \tag{12.129}
$$

$$
\ln\left(p_{1,-1}\right) \propto \sum_{k=1}^{N/3} \frac{r_{k+2N/3}h_k}{\sigma^2}, \tag{12.130}
$$

$$
\ln\left(p_{-1,1}\right) \propto \sum_{k=1}^{N/3} \frac{-r_{k+2N/3}h_k}{\sigma^2}, \tag{12.131}
$$

$$
\ln\left(p_{-1,-1}\right) \propto \sum_{k=1}^{N/3} \frac{-r_{k+N/3}h_k}{\sigma^2}. \tag{12.132}
$$

The expressions in (12.129) to (12.132) are the correlations of a reference impulse response $\pm\vec{h}$ with either the receive signal $r_{k+N/3}$ in the first time slot or $r_{k+2N/3}$ in the second time slot. Thus, these expressions correspond to the coherent receiver structure that can be expected in the case of full CSI. The maximum likelihood decision yields

$$[\hat{a}_1 = \hat{i}, \hat{a}_2 = \hat{j}] = \arg\max_{i,j} \ln(p_{i,j}), \tag{12.133}$$

where $\hat{a}_1$ and $\hat{a}_2$ are the estimated values for a_1 and a_2, respectively.

Instantaneous Power Delay Profile

For the next ML derivation a lower level of CSI is assumed. As noted in the derivations for the energy detector and the transmitted reference receiver, only the amplitudes are known in the case of knowledge of the instantaneous power delay profile. As described in section 12.2.2 the channel impulse response can be split in a part $\vec{x}$, which contains the magnitudes, and a part $\vec{z}$, which contains the signs, i.e., $\vec{h} = \vec{x} \odot \vec{z}$. $\odot$ denotes the element-wise multiplication. Hence, the CSI is given by $\mathcal{C}_{\mathrm{IPDP}} = \vec{x}$. Since the signs z_k are equiprobable, the probability density function is given by

$$p(z_k) = \frac{1}{2}\delta(z_k - 1) + \frac{1}{2}\delta(z_k + 1). \tag{12.134}$$

As in section 12.3.2, the signs of the CIRs have to be averaged out because they are unknown at the receiver side. With (12.123) this yields for the desired probability

$$p\left(\vec{r}|a_1 = a_i, a_2 = a_j, \mathcal{C}_{\mathrm{IPDP}}\right) = \prod_{k=1}^{N} \int_{-\infty}^{\infty} p(z_k) p\left(r_k|a_1 = a_i, a_2 = a_j, x_k, z_k\right) dz_k. \tag{12.135}$$

The probability for one receive sample r_k can be written as

$$p\left(r_k|a_1 = a_i, a_2 = a_j, x_k\right)$$

$$= \int_{-\infty}^{\infty} p(z_k)p\left(r_k|a_1 = a_i, a_2 = a_j, x_k, z_k\right) dz_k$$

$$= \frac{1}{2\sqrt{2\pi\sigma^2}} \cdot \exp\left\{-\frac{1}{2\sigma^2}\left(r_k^2 + 2x_k^2 - \left(2r_k + (a_1 + a_2)r_{k+N/3} + (a_1 - a_2)r_{k+2N/3}\right)x_k\right)\right\}$$

$$+ \frac{1}{2\sqrt{2\pi\sigma^2}} \cdot \exp\left\{-\frac{1}{2\sigma^2}\left(r_k^2 + 2x_k^2 + \left(2r_k + (a_1 + a_2)r_{k+N/3} + (a_1 - a_2)r_{k+2N/3}\right)x_k\right)\right\}.$$

$$(12.136)$$

By factoring out and removing

$$\frac{1}{2\sqrt{2\pi\sigma^2}} \cdot \exp\left\{-\frac{1}{2\sigma^2}\left(r_k^2 + 2x_k^2\right)\right\} \tag{12.137}$$

which is irrelevant for the the maximum likelihood decision one obtains

$$p\left(r_k|a_1 = a_i, a_2 = a_j, x_k\right)$$

$$\propto \exp\left\{+\frac{1}{2\sigma^2}\left(2r_k + (a_1 + a_2)r_{k+N/3} + (a_1 - a_2)r_{k+2N/3}\right)x_k\right\}$$

$$+ \exp\left\{-\frac{1}{2\sigma^2}\left(2r_k + (a_1 + a_2)r_{k+N/3} + (a_1 - a_2)r_{k+2N/3}\right)x_k\right\}$$

$$= 2\cosh\left(\frac{1}{2\sigma^2}\left(2r_k + (a_1 + a_2)r_{k+N/3} + (a_1 - a_2)r_{k+2N/3}\right)x_k\right). \tag{12.138}$$

This results in the following decision variables

$$p_{1,1} = p\left(\vec{r}|a_1 = 1, a_2 = 1, \vec{x}\right) \propto \prod_{k=1}^{N/3} \cosh\left(\frac{\left(r_k + r_{k+N/3}\right)x_k}{\sigma^2}\right), \tag{12.139}$$

$$p_{1,-1} = p\left(\vec{r}|a_1 = 1, a_2 = -1, \vec{x}\right) \propto \prod_{k=1}^{N/3} \cosh\left(\frac{\left(r_k + r_{k+2N/3}\right)x_k}{\sigma^2}\right), \tag{12.140}$$

$$p_{-1,1} = p\left(\vec{r}|a_1 = -1, a_2 = 1, \vec{x}\right) \propto \prod_{k=1}^{N/3} \cosh\left(\frac{\left(r_k - r_{k+2N/3}\right)x_k}{\sigma^2}\right), \tag{12.141}$$

and

$$p_{-1,-1} = p\left(\vec{r}|a_1 = -1, a_2 = -1, \vec{x}\right) \propto \prod_{k=1}^{N/3} \cosh\left(\frac{\left(r_k - r_{k+N/3}\right)x_k}{\sigma^2}\right) \tag{12.142}$$

for the different symbols. Taking the logarithms one gets

$$\ln(p_{1,1}) \propto \sum_{k=1}^{N/3} \ln\left(\cosh\left(\frac{(r_k + r_{k+N/3})\, x_k}{\sigma^2}\right)\right), \tag{12.143}$$

$$\ln(p_{1,-1}) \propto \sum_{k=1}^{N/3} \ln\left(\cosh\left(\frac{(r_k + r_{k+2N/3})\, x_k}{\sigma^2}\right)\right), \tag{12.144}$$

$$\ln(p_{-1,1}) \propto \sum_{k=1}^{N/3} \ln\left(\cosh\left(\frac{(r_k - r_{k+2N/3})\, x_k}{\sigma^2}\right)\right), \tag{12.145}$$

and

$$\ln(p_{-1,-1}) \propto \sum_{k=1}^{N/3} \ln\left(\cosh\left(\frac{(r_k - r_{k+N/3})\, x_k}{\sigma^2}\right)\right). \tag{12.146}$$

As noted for the energy detector in (12.27), the receive signal is correlated with the magnitude $\vec{x}$ of the channel impulse response. However, the reference pulse, which is also correlated with $\vec{x}$, is considered for the maximum likelihood decision, as well. Hence, the maximum likelihood receiver can be written as

$$[\hat{a}_1 = \hat{i}, \hat{a}_2 = \hat{j}] = \arg\max_{i,j} \ln(p_{i,j}). \tag{12.147}$$

Average Power Delay Profile

In the following the knowledge of the correlation matrix

$$\Lambda_{hh} = \begin{bmatrix} \mathcal{E}[h_1^2] & \cdots & 0 \\ \vdots & \ddots & \vdots \\ 0 & \cdots & \mathcal{E}[h_{N/2}^2] \end{bmatrix} = \begin{bmatrix} \lambda_{h,1} & \cdots & 0 \\ \vdots & \ddots & \vdots \\ 0 & \cdots & \lambda_{h,N/2} \end{bmatrix} \tag{12.148}$$

of $\vec{h}$ is assumed as explained in section 12.2.2. Again, the receive vector is rearranged such that the kth tap of $\vec{r}_1$, $\vec{r}_2$, and $\vec{r}_3$ are adjacent, i.e.,

$$\vec{r} = \left[r_1, r_{N/3+1}, r_{2N/3+1}, r_2, r_{N/3+2}, r_{2N/3+2}, \ldots, r_{N/3}, r_{2N/3}, r_N\right]. \tag{12.149}$$

Using this vector $\vec{r}$, the correlation matrix Λ_{rr} can be determined as

$$\Lambda_{rr} = \begin{bmatrix} \lambda_{r,11} & \lambda_{r,12} & \lambda_{r,13} & \ldots & 0 & 0 & 0 \\ \lambda_{r,21} & \lambda_{r,22} & 0 & \ldots & 0 & 0 & 0 \\ \lambda_{r,31} & 0 & \lambda_{r,33} & \ldots & 0 & 0 & 0 \\ \vdots & \vdots & \vdots & \ddots & \vdots & \vdots & \vdots \\ 0 & 0 & 0 & \ldots & \lambda_{r,N-2N-2} & \lambda_{r,N-2N-1} & \lambda_{r,N-2N} \\ 0 & 0 & 0 & \ldots & \lambda_{r,N-1N-2} & \lambda_{r,N-1N-1} & 0 \\ 0 & 0 & 0 & \ldots & \lambda_{r,NN-2} & 0 & \lambda_{r,NN} \end{bmatrix} \qquad (12.150)$$

The entries on the main diagonal of the correlation matrix Λ_{rr} are

$$\lambda_{r,kk} = \lambda_{h,\lceil k/3 \rceil} + \sigma^2 \qquad (12.151)$$

for k's that fulfill $(k \mod 3) = 1$,

$$\lambda_{r,kk} = \frac{1}{4}(a_1 + a_2)^2 \lambda_{h,\lceil k/3 \rceil} + \sigma^2 \qquad (12.152)$$

for k's that fulfill $(k \mod 3) = 2$, and

$$\lambda_{r,kk} = \frac{1}{4}(a_1 - a_2)^2 \lambda_{h,\lceil k/3 \rceil} + \sigma^2 \qquad (12.153)$$

for k's that fulfill $(k \mod 3) = 0$. As in the previous sections, $\lceil \cdot \rceil$ denotes a rounding up to the next higher integer number and $\mod$ denotes the rest of an integer division, i.e., the modulo operator. The remaining non-zero elements on the secondary diagonals of (12.150) are given at the positions, where $(k \mod 3) = 1$ by

$$\lambda_{r,kk+1} = \lambda_{r,k+1k} = \frac{1}{2}(a_1 + a_2)\lambda_{h,\lceil k/3 \rceil} \qquad (12.154)$$

and

$$\lambda_{r,kk+2} = \lambda_{r,k+2k} = \frac{1}{2}(a_1 - a_2)\lambda_{h,\lceil k/3 \rceil}. \qquad (12.155)$$

Knowing the correlation matrix Λ_{rr}, the desired probability can be calculated using [111]

$$p\left(\vec{r}|a_1 = a_i, a_2 = a_j, \Lambda_{rr}\right) = \left(\frac{1}{\sqrt{2\pi}}\right)^N \frac{1}{\sqrt{\Delta_{rr}}} \exp\left\{-\frac{1}{2}\vec{r}^T \Lambda_{rr}^{-1} \vec{r}\right\} \qquad (12.156)$$

with the determinant Δ_{rr} of Λ_{rr} and the transpose $\vec{r}^T$ of $\vec{r}$. The determinant Δ_{rr} can be obtained as

$$\Delta_{rr} = \prod_{k=1}^{N/2} 2\lambda_{h,k}\sigma^4 + \sigma^6. \tag{12.157}$$

Since Δ_{rr} is independent of the symbols a_1 and a_2, it can be neglected for the derivation of the maximum likelihood receiver. The inverse of the correlation matrix Λ_{rr}^{-1} has non-zero elements at the same positions as the correlation matrix Λ_{rr} in (12.150). The elements on the main diagonal are given by

$$\iota_{r,kk} = \frac{\lambda_{h,\lceil k/3 \rceil} + \sigma^2}{\sigma^2 \left(2\lambda_{h,\lceil k/3 \rceil} + \sigma^2\right)} \tag{12.158}$$

if $(k \mod 3) = 1$,

$$\iota_{r,kk} = \frac{\left(4 + (a_1 - a_2)^2\right)\lambda_{h,\lceil k/3 \rceil} + 4\sigma^2}{4\sigma^2 \left(2\lambda_{h,\lceil k/3 \rceil} + \sigma^2\right)} \tag{12.159}$$

if $(k \mod 3) = 2$, and

$$\iota_{r,kk} = \frac{\left(4 + (a_1 + a_2)^2\right)\lambda_{h,\lceil k/3 \rceil} + 4\sigma^2}{4\sigma^2 \left(2\lambda_{h,\lceil k/3 \rceil} + \sigma^2\right)} \tag{12.160}$$

if $(k \mod 3) = 0$. The remaining elements are non-zero only for $(k \mod 3) = 1$ and can be determined as

$$\iota_{r,kk+1} = \iota_{r,k+1k} = -\frac{(a_1 + a_2)\lambda_{h,\lceil k/3 \rceil}\left((a_1 - a_2)^2\lambda_{h,\lceil k/3 \rceil} + 4\sigma^2\right)}{8\sigma^4 \left(2\lambda_{h,\lceil k/3 \rceil} + \sigma^2\right)} \tag{12.161}$$

and

$$\iota_{r,kk+2} = \iota_{r,k+2k} = -\frac{(a_1 - a_2)\lambda_{h,\lceil k/3 \rceil}\left((a_1 + a_2)^2\lambda_{h,\lceil k/3 \rceil} + 4\sigma^2\right)}{8\sigma^4 \left(2\lambda_{h,\lceil k/3 \rceil} + \sigma^2\right)}. \tag{12.162}$$

By inserting (12.157) - (12.162) into (12.156) and by neglecting the pre-factor one obtains for the four different hypotheses

$$p\left(\vec{r}\,|a_1 = 1, a_2 = 1, \Lambda_{rr}\right)$$

$$\propto \exp\left\{-\frac{1}{2}\sum_{k=1}^{N/3}\frac{\left((r_k - r_{k+N/3})^2 + 2r_{k+2N/3}^2\right)\lambda_{h,k} + \left(r_k^2 + r_{k+N/3}^2 + r_{k+2N/3}^2\right)\sigma^2}{\sigma^2\left(2\lambda_{h,k} + \sigma^2\right)}\right\},$$

(12.163)

$$p\left(\vec{r}\,|a_1 = 1, a_2 = -1, \Lambda_{rr}\right)$$

$$\propto \exp\left\{-2\sum_{k=1}^{N/3}\frac{\left((r_k - r_{k+2N/3})^2 + 2r_{k+N/3}^2\right)\lambda_{h,k} + \left(r_k^2 + r_{k+N/3}^2 + r_{k+2N/3}^2\right)\sigma^2}{\sigma^2\left(2\lambda_{h,k} + \sigma^2\right)}\right\},$$

(12.164)

$$p\left(\vec{r}\,|a_1 = -1, a_2 = 1, \Lambda_{rr}\right)$$

$$\propto \exp\left\{-2\sum_{k=1}^{N/3}\frac{\left((r_k + r_{k+2N/3})^2 + 2r_{k+N/3}^2\right)\lambda_{h,k} + \left(r_k^2 + r_{k+N/3}^2 + r_{k+2N/3}^2\right)\sigma^2}{\sigma^2\left(2\lambda_{h,k} + \sigma^2\right)}\right\},$$

(12.165)

$$p\left(\vec{r}\,|a_1 = -1, a_2 = -1, \Lambda_{rr}\right)$$

$$\propto \exp\left\{-2\sum_{k=1}^{N/3}\frac{\left((r_k + r_{k+N/3})^2 + 2r_{k+2N/3}^2\right)\lambda_{h,k} + \left(r_k^2 + r_{k+N/3}^2 + r_{k+2N/3}^2\right)\sigma^2}{\sigma^2\left(2\lambda_{h,k} + \sigma^2\right)}\right\}.$$

(12.166)

The term

$$\frac{(r_k^2 + r_{k+N/3}^2 + r_{k+2N/3}^2)\sigma^2}{\sigma^2\left(2\lambda_{h,k} + \sigma^2\right)}$$

(12.167)

is the same for all four hypotheses and can be omitted.

Therefore, the probabilities which are relevant for the maximum likelihood decision are given by

$$p_{1,1} = p\left(\vec{r}|a_1 = 1, a_2 = 1, \Lambda_{rr}\right) \propto \exp\left\{-\sum_{k=1}^{N/3} \frac{(r_k - r_{k+N/3})^2 + 2r_{k+2N/3}^2}{2\sigma^2\left(2 + \frac{\sigma^2}{\lambda_{h,k}}\right)}\right\}, \quad (12.168)$$

$$p_{1,-1} = p\left(\vec{r}|a_1 = 1, a_2 = -1, \Lambda_{rr}\right) \propto \exp\left\{-\sum_{k=1}^{N/3} \frac{(r_k - r_{k+2N/3})^2 + 2r_{k+N/3}^2}{2\sigma^2\left(2 + \frac{\sigma^2}{\lambda_{h,k}}\right)}\right\},$$

$$(12.169)$$

$$p_{-1,1} = p\left(\vec{r}|a_1 = -1, a_2 = 1, \Lambda_{rr}\right) \propto \exp\left\{-\sum_{k=1}^{N/3} \frac{(r_k + r_{k+2N/3})^2 + 2r_{k+N/3}^2}{2\sigma^2\left(2 + \frac{\sigma^2}{\lambda_{h,k}}\right)}\right\},$$

$$(12.170)$$

$$p_{-1,-1} = p\left(\vec{r}|a_1 = -1, a_2 = -1, \Lambda_{rr}\right) \propto \exp\left\{-\sum_{k=1}^{N/3} \frac{(r_k + r_{k+N/3})^2 + 2r_{k+2N/3}^2}{2\sigma^2\left(2 + \frac{\sigma^2}{\lambda_{h,k}}\right)}\right\},$$

$$(12.171)$$

where the maximum out of (12.168) to (12.171) determines the transmitted symbols a_1 and a_2. Using the logarithm, which is strictly monotonic increasing, the decision variables in (12.168) to (12.171) can be rewritten such that the minimum of the following four hypotheses yields the transmitted symbols

$$\ln(p_{1,1}) \propto \sum_{k=1}^{N/3} \frac{(r_k - r_{k+N/3})^2 + 2r_{k+2N/3}^2}{2\sigma^2\left(2 + \frac{\sigma^2}{\lambda_{h,k}}\right)}$$

$$= \frac{1}{2\sigma^2} \sum_{k=1}^{N/3} \frac{-2r_k r_{k+N/3} + r_{k+2N/3}^2}{2 + \frac{\sigma^2}{\lambda_{h,k}}}, \quad (12.172)$$

$$\ln(p_{1,-1}) \propto \sum_{k=1}^{N/3} \frac{(r_k - r_{k+2N/3})^2 + 2r_{k+N/3}^2}{2\sigma^2\left(2 + \frac{\sigma^2}{\lambda_{h,k}}\right)}$$

$$= \frac{1}{2\sigma^2} \sum_{k=1}^{N/3} \frac{-2r_k r_{k+2N/3} + r_{k+N/3}^2}{2 + \frac{\sigma^2}{\lambda_{h,k}}}, \quad (12.173)$$

$$\ln(p_{-1,1}) \propto \sum_{k=1}^{N/3} \frac{\left(r_k + r_{k+2N/3}\right)^2 + 2r_{k+N/3}^2}{2\sigma^2 \left(2 + \frac{\sigma^2}{\lambda_{h,k}}\right)}$$

$$= \frac{1}{2\sigma^2} \sum_{k=1}^{N/3} \frac{2r_k r_{k+2N/3} + r_{k+N/3}^2}{2 + \frac{\sigma^2}{\lambda_{h,k}}}, \tag{12.174}$$

$$\ln(p_{-1,-1}) \propto \sum_{k=1}^{N/3} \frac{\left(r_k + r_{k+N/3}\right)^2 + 2r_{k+2N/3}^2}{2\sigma^2 \left(2 + \frac{\sigma^2}{\lambda_{h,k}}\right)}$$

$$= \frac{1}{2\sigma^2} \sum_{k=1}^{N/3} \frac{2r_k r_{k+N/3} + r_{k+2N/3}^2}{2 + \frac{\sigma^2}{\lambda_{h,k}}}. \tag{12.175}$$

From (12.172) to (12.175) it can be observed that the first term in the numerator tends towards 0 for the effectively transmitted symbols while all other terms are growing. In the case of knowledge of the APDP the maximum likelihood receiver is given by

$$[\hat{a}_1 = \hat{i}, \hat{a}_2 = \hat{j}] = \arg \min_{i,j} \ln(p_{i,j}). \tag{12.176}$$

No Channel State Information

Finally, it is assumed that no channel state information is available at the receiver side. However, to calculate the maximum likelihood receiver it is assumed that the channel taps are Gaussian distributed according to

$$p(h_k) = \frac{1}{\sqrt{2\pi\sigma_h^2}} \exp\left\{-\frac{h_k^2}{2\sigma_h^2}\right\}. \tag{12.177}$$

The desired probability can be calculated by averaging over the channel distribution, i.e.,

$$p\left(\vec{r}|a_1 = a_i, a_2 = a_j\right) = \prod_{k=1}^{N} \int_{-\infty}^{\infty} p(h_k)p\left(r_k|a_1 = a_i, a_2 = a_j, h_k\right) dh_k \tag{12.178}$$

with $p\left(r_k | a_1 = a_i, a_2 = a_j, h_k\right)$ from (12.123). By evaluating the integral in (12.178) one gets

$$
\begin{aligned}
p\left(\vec{r} | a_1 = a_i, a_2 = a_j\right) = \prod_{k=1}^{N/3} & \frac{1}{\sqrt{2\pi(\sigma^2 + 2\sigma_h^2)}} \\
\cdot \exp & \left\{ \frac{4r_k \left(a_2(r_{k+N/3} - r_{k+2N/3}) + a_1(r_{k+N/3} + r_{k+2N/3})\right)\sigma_h^2}{8\sigma^2 \left(\sigma^2 + 2\sigma_h^2\right)} \right\} \\
\cdot \exp & \left\{ \frac{\left(a_2(r_{k+N/3} - r_{k+2N/3}) + a_1(r_{k+N/3} + r_{k+2N/3})\right)^2 \sigma_h^2}{8\sigma^2 \left(\sigma^2 + 2\sigma_h^2\right)} \right\} \\
\cdot \exp & \left\{ \frac{-4r_k^2(\sigma^2 + \sigma_h^2)}{8\sigma^2 \left(\sigma^2 + 2\sigma_h^2\right)} \right\} .
\end{aligned}
\tag{12.179}
$$

After inserting the different symbol combinations into (12.179) the probabilities of the four hypotheses are given by

$$
p_{1,1} = p\left(\vec{r} | a_1 = 1, a_2 = 1\right) = \prod_{k=1}^{N/3} \frac{\exp\left\{ \frac{2r_k r_{k+N/3}\sigma_h^2 + r_{k+N/3}^2\sigma_h^2 - r_k^2(\sigma^2 + \sigma_h^2)}{2\sigma^2(\sigma^2 + 2\sigma_h^2)} \right\}}{\sqrt{2\pi(\sigma^2 + 2\sigma_h^2)}},
\tag{12.180}
$$

$$
p_{1,-1} = p\left(\vec{r} | a_1 = 1, a_2 = -1\right) = \prod_{k=1}^{N/3} \frac{\exp\left\{ \frac{2r_k r_{k+2N/3}\sigma_h^2 + r_{k+2N/3}^2\sigma_h^2 - r_k^2(\sigma^2 + \sigma_h^2)}{2\sigma^2(\sigma^2 + 2\sigma_h^2)} \right\}}{\sqrt{2\pi(\sigma^2 + 2\sigma_h^2)}},
\tag{12.181}
$$

$$
p_{-1,1} = p\left(\vec{r} | a_1 = -1, a_2 = 1\right) = \prod_{k=1}^{N/3} \frac{\exp\left\{ \frac{-2r_k r_{k+2N/3}\sigma_h^2 + r_{k+2N/3}^2\sigma_h^2 - r_k^2(\sigma^2 + \sigma_h^2)}{2\sigma^2(\sigma^2 + 2\sigma_h^2)} \right\}}{\sqrt{2\pi(\sigma^2 + 2\sigma_h^2)}},
\tag{12.182}
$$

$$
p_{-1,-1} = p\left(\vec{r} | a_1 = -1, a_2 = -1\right) = \prod_{k=1}^{N/3} \frac{\exp\left\{ \frac{-2r_k r_{k+N/3}\sigma_h^2 + r_{k+N/3}^2\sigma_h^2 - r_k^2(\sigma^2 + \sigma_h^2)}{2\sigma^2(\sigma^2 + 2\sigma_h^2)} \right\}}{\sqrt{2\pi(\sigma^2 + 2\sigma_h^2)}}.
\tag{12.183}
$$

Similar to the previous derivations the variables for the maximum likelihood decision can be simplified by taking the logarithm and omitting the denominators as well as the last term in the numerator of the exponent, as these are the same for all hypotheses. Thus, the transmitted symbols can be determined by finding the maximum out of

$$\ln(p_{1,1}) \propto \sum_{k=1}^{N/3} 2r_k r_{k+N/3} + r_{k+N/3}^2, \tag{12.184}$$

$$\ln(p_{1,-1}) \propto \sum_{k=1}^{N/3} 2r_k r_{k+2N/3} + r_{k+2N/3}^2, \tag{12.185}$$

$$\ln(p_{-1,1}) \propto \sum_{k=1}^{N/3} -2r_k r_{k+2N/3} + r_{k+2N/3}^2, \tag{12.186}$$

$$\ln(p_{-1,-1}) \propto \sum_{k=1}^{N/3} -2r_k r_{k+N/3} + r_{k+N/3}^2, \tag{12.187}$$

i.e., the maximum likelihood decision is made according to

$$[\hat{a}_1 = \hat{i}, \hat{a}_2 = \hat{j}] = \arg\max_{i,j} \ln(p_{i,j}). \tag{12.188}$$

The resulting decision variables in (12.184) to (12.187) are an interesting construction. The first terms are the correlations of the reference pulse with the expected position of the data pulse corresponding to the transmit symbols. The second terms are the energies at the expected position of the data pulse. Hence, the resulting receiver structure for the TR PIAM is a combination of the transmitted reference receiver and the energy detector.

12.5.2 Performance Results

As in the cases of binary PPM and TR PAM the different receiver structures are compared by means of their BER curves. The channel taps are assumed to be uniformly distributed within the duration of the channel impulse response as described in section 12.2.3. Moreover, for the simulations only the no ISI case is considered, i.e., the CIR durations are 10 ns. In Fig. 12.11, the bit error curves are shown for the TR PIAM ML receiver structures with different CSI. It can be seen that the the ML with full CSI yields the best performance, followed by the ML with knowledge of the IPDP. The performance of the MLs with knowledge of the APDP and without knowledge are the same. This behavior is caused by the uniformly distributed channel taps. For such a channel, the decision rule for the ML with APDP knowledge reduces to the decision

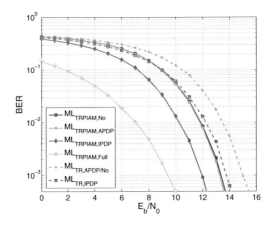

Figure 12.11: Bit error rate curves for TRPIAM in the case of no ISI

rule for the ML without CSI. Additionally, except from the full CSI case the performance of the TR PIAM ML receivers is about 2 dB better compared to the performance of the TR PAM ML receivers.

12.6 Suboptimum Receiver Structures

In the following suboptimum receiver structures for binary pulse position modulation, transmitted reference pulse amplitude modulation, and transmitted reference pulse interval amplitude modulation are presented. For the derivations the non-ISI case is considered. The performance loss of the suboptimum receiver structures is compared to the optimum receiver structures by means of simulations.

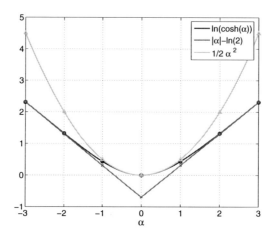

Figure 12.12: $\ln\left(\cosh(\alpha)\right)$ and its approximations

12.6.1 Suboptimum Receiver Structures for Binary PPM

The binary PPM receiver knowing the IPDP requires a $\ln\left((\cosh(\alpha)\right)$ operation. This operation can be approximated by [28]

$$\ln\left(\cosh(\alpha)\right) \approx \ln\left(e^{|\alpha|}\right) - \ln(2) = |\alpha| - \ln(2) \tag{12.189}$$

for $\alpha > 2$. For small α there can be another approximation of $\ln(\cosh(\alpha))$ provided. Assuming

$$\cosh(\alpha) \approx 1 + \frac{1}{2}\alpha^2 \tag{12.190}$$

and

$$\ln(1 + y) \approx y \tag{12.191}$$

yields the approximation

$$\ln(\cosh(\alpha)) \approx \frac{1}{2}\alpha^2. \tag{12.192}$$

The operation $\ln\left((\cosh(\alpha)\right)$ and its approximations are shown in Fig. 12.12.

Based on the ML decision

$$L = \sum_{k=1}^{N/2} \left(\ln \left[\cosh \left(\frac{r_{k+N/2} x_k}{\sigma^2} \right) \right] - \ln \left[\cosh \left(\frac{r_k x_k}{\sigma^2} \right) \right] \right) \qquad (12.193)$$

the resulting approximations for the log-likelihood are

$$L = \sum_{k=1}^{N/2} \left| \frac{r_{k+N/2} x_k}{\sigma^2} \right| - \left| \frac{r_k x_k}{\sigma^2} \right| \qquad (12.194)$$

using (12.189) and

$$L = \frac{1}{2} \sum_{k=1}^{N/2} \left(\frac{r_{k+N/2} x_k}{\sigma^2} \right)^2 - \left(\frac{r_k x_k}{\sigma^2} \right)^2 \qquad (12.195)$$

using (12.192).

12.6.2 Suboptimum Receiver Structures for TR PAM

Similar to the binary PPM, the hyperbolic cosine is also required for the TR PAM in the case of the IPDP knowledge. Therefore, the approximations from (12.189) and (12.192) can also be applied to the decision

$$L = \sum_{k=1}^{N/2} \ln \left(\cosh \left(\frac{(r_k + r_{k+N/2}) x_k}{\sigma^2} \right) \right) - \ln \left(\cosh \left(\frac{(r_k - r_{k+N/2}) x_k}{\sigma^2} \right) \right). \qquad (12.196)$$

Using these approximations yields the expressions

$$L = \sum_{k=1}^{N/2} \left| \frac{(r_k + r_{k+N/2}) x_k}{\sigma^2} \right| - \left| \frac{(r_k - r_{k+N/2}) x_k}{\sigma^2} \right| \qquad (12.197)$$

for large arguments and

$$L = \frac{1}{2} \sum_{k=1}^{N/2} \left(\frac{(r_k + r_{k+N/2}) x_k}{\sigma^2} \right)^2 - \left(\frac{(r_k - r_{k+N/2}) x_k}{\sigma^2} \right)^2 \qquad (12.198)$$

for small arguments.

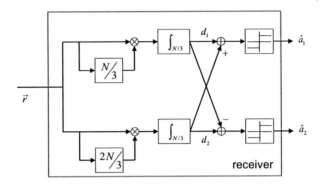

Figure 12.13: Suboptimum receiver structure for TR PIAM

12.6.3 Suboptimum Receiver Structures for TR PIAM without CSI

It could be seen from (12.184) to (12.187) that the decision variables for TR PIAM combine a transmitted reference receiver and an energy detector in the case of no CSI. Since the decision can be made by using only the transmitted reference part, a suboptimum receiver structure can be obtained by omitting the energy detector part [22]. Such a receiver with different delays $T_1 = T_I$ and $T_2 = T_I + T_D$ is shown in Fig. 12.13. Depending on the transmit signal the output signal of one correlator is the correlation of the reference signal with the information signal while the output signal of the other correlator is the correlation of the reference signal with noise.

The output signals r_1 and r_2 of the correlators are fed into two adders. Behind the adders the transmit signal is estimated as

$$
\hat{a}_1 = \begin{cases} -1 & \text{if} \quad d_1 + d_2 < 0 \\ 1 & \text{otherwise} \end{cases} \tag{12.199}
$$

and

$$
\hat{a}_2 = \begin{cases} -1 & \text{if} \quad d_1 - d_2 < 0 \\ 1 & \text{otherwise} \end{cases}
$$

using two slicers.

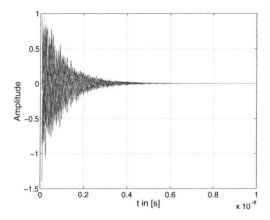

Figure 12.14: Exemplary channel impulse responses with an exponential decay

12.7 Comparison of the Different Receiver Structures

In the previous sections the maximum likelihood and some suboptimal receiver structures for binary PPM, TR PAM, and TR PIAM have been derived. These receiver structures are compared by means of bit error simulations in the following. The BERs are plotted over the signal-to-noise ratio E_b/N_0, where E_b denotes the energy per bit and $N_0/2$ is the noise power spectral density. The receiver structures have an integration time of 10 ns and ISI does not occur. For the simulations, three different kinds of channels are considered.

First, a channel with uniformly distributed channel taps is assumed. The duration of the channel is 10 ns, which corresponds with the integration duration of the receivers. Since the receiver with APDP knowledge equals the receiver without CSI for such a channel an exponential decaying channel with Gaussian distributed taps is also considered. In Fig. 12.14, channel impulse responses with exponential decay used for the simulations are shown exemplarily. Besides these both different channel types also the BAN channel model derived in section 7 is used for the receiver comparison. There, the channel taps are decaying even faster than in the exponential decaying channel.

In Fig. 12.15 the BER curves for binary PPM are shown for the different receiver structures.

179

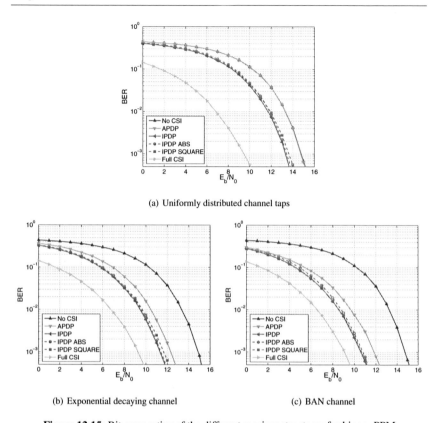

(a) Uniformly distributed channel taps

(b) Exponential decaying channel

(c) BAN channel

Figure 12.15: Bit error ratios of the different receiver structures for binary PPM

It can be seen that the receiver with full CSI and the one with no CSI, i.e., the energy detector, have the same performance independently of the channel characteristics. For the energy detector the performance only depends on the energy contained in the channel as well as on the integration duration. This observation corresponds with the results in [113]. The receivers with knowledge of the IPDP or APDP show a different performance for the different channels depending on the power delay profile. The biggest performance difference can be observed for the APDP. In the case of the uniformly distributed channel taps the receiver equals the energy detector. For the other channels, the main part of the energy is contained only in a fraction of the channel. The receiver, which knows the APDP, considers this information by using only these parts of the channel impulse response and blanking the remaining taps. Hence, the performance improves compared to the energy detector for the channel taps, which are not uniformly distributed. For the BAN channel model the performance of the receiver with APDP knowledge is only slightly worse compared to receiver with IPDP knowledge. In the case of the decaying channels this receiver structure is only about 2 dB worse than the receiver with full CSI. The receivers using the absolute value, i.e., IPDP ABS, and the squaring, i.e., IDPD SQUARE, for the approximation of the IPDP receiver as presented in (12.194) and (12.195) exhibit a very good performance. In particular, the IPDP ABS, which uses the approximation for large arguments, fits the BER curve of the IPDP receiver for $E_b/N_0 > 10$ dB. Below $E_b/N_0 \approx 10$ dB the IPDP SQUARE curve is better than the one for IPDP ABS and corresponds with the BER curve of the IPDP receiver. However, with increasing E_b/N_0 the BER curves for IPDP and IPDP SQUARE diverge. This can be expected due to the fact that the IPDP SQUARE is based on an approximation that is valid for small arguments.

The BER curves for the transmitted reference PAM receivers are shown in Fig. 12.16 for the different channels. It can be seen that the BER curves of the different receiver structures are the same for the different channels as the ones for the binary PPM. The performance equality of the transmitted reference receiver and energy detector, i.e, the receiver structures without CSI, is shown in [84]. However, from the BER curves in Fig. 12.15 and Fig. 12.16 it can be seen that this equality is also given for the receiver structures that know the IPDP and the APDP. Hence, the same conclusions as for the binary PPM can be made. In particular for the exponential decaying channel and the BAN channel, the receiver with APDP knowledge shows almost the same performance as the one with IPDP knowledge. Since the estimation of the APDP is not as complex as the estimation of the IPDP, the receiver with APDP knowledge is preferred to the receiver with IPDP knowledge.

Compared to the binary PPM and the TR PAM, most TR PIAM receivers have a better BER performance as it can be seen in Fig. 12.17. Only the coherent receiver shows the same performance as the coherent receivers for binary PPM and TR PAM. In the case of no CSI the TR PIAM receivers are independent of the channel about 1 dB better than the TR PAM and PPM

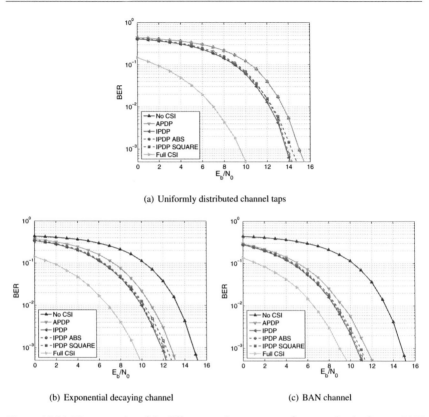

(a) Uniformly distributed channel taps

(b) Exponential decaying channel

(c) BAN channel

Figure 12.16: Bit error ratios of the different receiver structures for transmitted reference PAM

receivers. The suboptimum receiver structure for TR PIAM, which is presented in Fig. 12.13, yields almost the same performance as the ML receiver without CSI. Hence, the energy detector parts in the ML receiver can be skipped with loosing only marginal performance. The performance of the TR PIAM receivers with knowledge of APDP and IPDP is about 2 dB better compared to the receivers for TR PAM and binary PPM with corresponding CSI. For the exponential decaying channel, where the energy is condensed, the receiver with IPDP knowledge is only about 1 dB worse than the receiver with full CSI and the one with APDP knowledge only about 2 dB, respectively. For the BAN channel the gap between the BER curves even decreases.

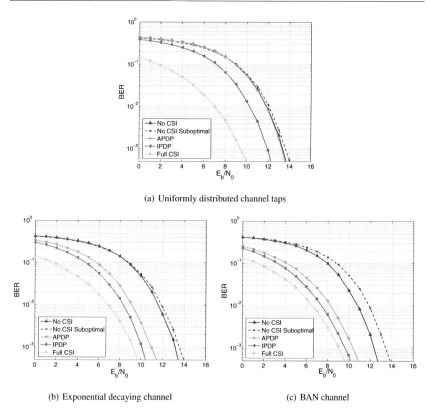

(a) Uniformly distributed channel taps

(b) Exponential decaying channel

(c) BAN channel

Figure 12.17: Bit error ratios of the different receiver structures for transmitted reference PIAM

The gap between receivers with full CSI and IPDP is about 0.5 dB for BER$= 10^{-3}$. However, the receiver with APDP knowledge is more attractive, since it is only about half an additional dB worse and it allows for a simpler acquisition of CSI.

12.7.1 Discussion of the Performance Results

Full CSI   From Fig. 12.15 to Fig. 12.17 it could have been seen that all three receiver structures having full CSI show the same bit error performance. In the case of full CSI, coherent

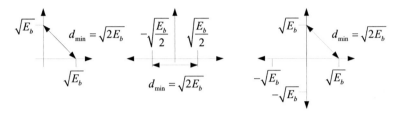

Figure 12.18: Signal space representations and minimum distances for the different modulation schemes

detection is possible and the performance of the receiver structures is given by the minimum distances between the different symbols. The signal space representation and the minimum distances for the three modulation schemes are presented in Fig. 12.18. The signal space representation for binary PPM, which is an orthogonal modulation scheme, is shown on the left side of Fig. 12.18. Assuming an energy per bit E_b the minimum distance between the two possible symbols is given by the distance of the data pulses $d_{min} = \sqrt{2E_b}$. Using transmitted reference PAM, which is an antipodal modulation scheme, the energy per bit E_b for decision is reduced by factor 2, because the reference pulse does not contain information. Hence, the minimum distance is also given by $d_{min} = \sqrt{2E_b}$, which can be seen from the signal space representation in the middle of Fig. 12.18. The energy loss due to the reference pulse is compensated in TR PIAM by the transmission of two bits per symbol. Due to the combination of orthogonal and antipodal modulation the signal space representation is also a combination of the corresponding signal space representations. The energy per bit E_b is the same as for the binary PPM. The signal space representation for TR PIAM is presented on the right side of Fig. 12.18. The same as for binary PPM and the transmitted reference PAM, the minimum distance for TR PIAM is also $d_{min} = \sqrt{2E_b}$. This result verifies the same performance of the coherent receivers for the three different modulation schemes.

IPDP The ML estimators for the three modulation schemes having knowledge of the IPDP are given in (12.27), (12.64), and(12.147). Assuming the high SNR approximation from (12.189) the decision metrics are given by

$$L = \sum_{k=1}^{N/2} \left(\left| r_{k+N/2} \right| - \left| r_k \right| \right) \frac{x_k}{\sigma^2} \qquad (12.200)$$

for binary PPM,

$$L = \sum_{k=1}^{N/2} \left(\left| r_k + r_{k+N/2} \right| - \left| r_k - r_{k+N/2} \right| \right) \frac{x_k}{\sigma^2} \qquad (12.201)$$

for TR PAM, and

$$L = \max_{i,j} \left\{ \sum_{k=1}^{N/3} \left| r_k + r_{k+N/3} \right| \frac{x_k}{\sigma^2}; \sum_{k=1}^{N/3} \left| r_k + r_{k+2N/3} \right| \frac{x_k}{\sigma^2}; \right.$$
$$\left. \sum_{k=1}^{N/3} \left| r_k - r_{k+2N/3} \right| \frac{x_k}{\sigma^2}; \sum_{k=1}^{N/3} \left| r_k - r_{k+N/3} \right| \frac{x_k}{\sigma^2} \right\} \qquad (12.202)$$

for TR PIAM. In all three decision metrics a weighting with $\frac{x_k}{\sigma^2}$, i.e., the ratio of instantaneous amplitude and noise variance, is performed. Hence, channel taps are skipped, where the noise variance is much larger than the instantaneous channel tap amplitude. The performance for binary PPM and TR PAM is the same since the bit energy, which is reduced by the factor 2, has to be considered again for TR PAM. Thereby, the distance between the two possible symbols is the same in (12.200) and (12.201). For TR PIAM the distance in (12.202) is larger, since the energy per bit is the same as for binary PPM. However, the performance gain of TR PIAM is less than 3 dB compared to the both other modulation schemes, since noise from three instead of two time slots has to be considered for the TR PIAM ML decision.

APDP The ML estimators in the case of APDP are given in (12.42), (12.80), and (12.176) for binary PPM, TR PAM, and TR PIAM, respectively. Likewise in the case of IPDP, a weighting is done by these receiver structures. However, the average per tap SNR is considered here, i.e., if the average channel tap amplitude λ_k is small compared to the noise variance σ^2, the corresponding receive signal tap is not considered for decision. Hence, the receiver structures with IPDP or APDP knowledge improve their performance for channels with decaying power delay profiles. There, channel taps that contain no or only marginal energy contributions are omitted for decision. Again, the TR PIAM performs better than both other modulation schemes. The reason is the same as in the case of IPDP. Additionally, the decision rule in (12.176) combines also a transmitted reference and an energy detector part.

No CSI   In [84] it has been shown that in the case of no CSI the performance for binary PPM and TR PAM is the same. There, it has been shown by using a Gaussian approximation of the noise that the decision SNR for both receiver structures is the same and hence yields the same performance. Alike in the both previous cases of IPDP and APDP knowledge, the TR PIAM modulation scheme does not exhibit the E_b penalty of the TR PAM receiver. However, the presence of noise in three time slots has to be considered here, too. The performance of the three receiver structures without CSI mainly depends on the integration time. If the integration time is longer than the part of the channel containing the substantial part of the energy, mainly noise is collected during this additional integration time. However, integration time which is too short results in omitting a substantial part of the energy that would improve the performance. Hence, there exists an optimum integration time (cf. [84]).

Integration Duration   As mentioned above, the integration time has a severe impact on the performance of the presented receiver structures. Therefore, BER plots are shown in the following with respect to the integration duration. The BER curves are plotted for the two values $E_b/N_0 = 10$ dB and $E_b/N_0 = 12$ dB.

In Fig. 12.19, the BER curves are presented for a channel with uniformly distributed channel taps. Different from the previous section, the length of the channel is only 5 ns. For the receiver structure without CSI the bit error performance improves up to this point. By increasing the integration duration above 5 ns the performance decreases, because only noise is collected during this time. The other receivers show a different behavior. Although their performance improves up to an integration time of 5 ns, as well, the performance does not get worse with increasing integration time above this threshold. These receiver structures suppress the parts of the receive signal, where only noise is contained and they do not consider it for decision. This shows that the receiver without CSI is more sensitive to an inappropriately chosen integration duration than

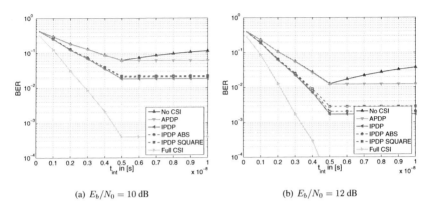

(a) $E_b/N_0 = 10$ dB

(b) $E_b/N_0 = 12$ dB

Figure 12.19: Impact of the integration duration on the bit error ratios of the different receiver structures for binary PPM using uniformly distributed channel taps

the receiver structures with higher CSI level. Moreover, it can be observed from the curves in Fig. 12.19 that due to the weighting of the receive taps the receivers with full CSI and IPDP knowledge benefit more from each additional sample considered for decision than the receivers without CSI or with APDP knowledge. This causes the different slopes of the BER curves in the first 5 ns. Besides the level of CSI knowledge also the signal-to-noise ratio E_b/N_0 has an impact on the slope. By having a higher E_b/N_0 the slopes of the BER curves become steeper and the optimum bit error rates get better.

The bit error curves for the exponential channel, which was presented in the previous section, are shown in Fig. 12.20. Since this channel does not contain such a step as the previously considered channel, the bit error curves show a different behavior. It can be observed that the optimum integration time for the receiver without CSI is about $1.5 - 2$ ns. Again, the performance decreases the further the integration duration increases. For the remaining receiver structures with higher level of CSI the optimum integration time is not the same as for the receiver without CSI. It can be seen in Fig. 12.20 that the optimum integration time increases with increasing CSI. While the receiver with IPDP knowledge has an optimum integration time of about 3 ns the receiver with full CSI improves its performance by considering channel taps

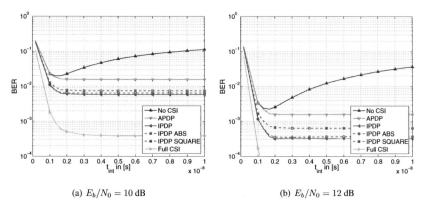

(a) $E_b/N_0 = 10$ dB

(b) $E_b/N_0 = 12$ dB

Figure 12.20: Impact of the integration duration on the bit error ratios of the different receiver structures for binary PPM using exponentially decaying channel taps

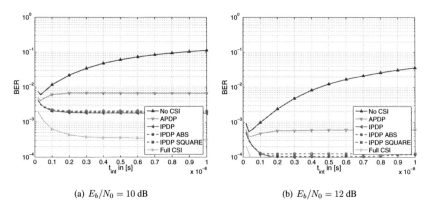

(a) $E_b/N_0 = 10$ dB

(b) $E_b/N_0 = 12$ dB

Figure 12.21: Impact of the integration duration on the bit error ratios of the different receiver structures for binary PPM using the BAN channel model

up to about 4 ns. As already noted in the previous case we observe again that the receivers with higher CSI level are less sensitive to the integration time than the receiver without CSI.

In Fig. 12.21, the bit error curves are shown for the BAN channel model derived in section 7. The trends for the different curves are similar to the ones in Fig. 12.20. However, since the energy is contained in a very short time window for this kind of channels, the optimum integration time is smaller than the one for the previously considered channels. For the receiver without CSI it can be observed that a slight increase of the integration time above the optimum point results in a moderate performance loss while a slight decrease of the integration time influences the performance severely. As already shown in Fig. 12.19 and Fig. 12.20 the receivers with higher CSI level are robust against the increase of the integration duration due to the suppressing of channel taps that contain mainly noise parts. Different from the other curves, a slight performance decrease of the APDP receiver can be observed for integration durations larger than the optimum one. This decrease of the BER is caused by the following. Since the receiver with APDP knowledge uses the average distribution of the channel taps for the weighting of the receive signal, it can happen that some noisy receive taps at the beginning of the channel impulse response are weighted with a large weight. Moreover, it can be observed that the optimum integration duration of the energy detector increases with increasing E_b/N_0. With higher E_b/N_0, i.e., smaller noise contributions, also some small channel taps increase the decision performance.

Part V

Medium Access Control

Chapter 13

Temporal Cognitive Medium Access

Since UWB coexistence and interference issues are very important, there have been already issued several publications in this area. Up to now, most investigations of coexistence issues have concerned the interference of UWB devices on existing services, such as [114], [115], [116], [117] . There are also some publications which consider the impact of the existing system's interference on UWB systems, such as [118], [119] , [120]. But there are only few publications, such as [121] , [122] , [19], in which the interference mitigation techniques are considered.

General interference mitigation methods, which are not limited to UWB only, are presented in [123]. There, collaborative and non-collaborative coexistence mechanisms are proposed. In the collaborative scenario different wireless systems are able to share information and to negotiate channel access. In the non-collaborative scenario different systems do not have the ability to coordinate their transmission. There, wireless systems can only use strategies such as carrier sense multiple access (CSMA) or adaptive frequency hopping. The disadvantage of these strategies is that maximum efficiency of the channel is not used. However, since the existing services are usually not collaborating, we consider the non-collaborative approach as the more promising one for UWB systems and we use it as the basis for our considerations.

The IEEE 802.15 subgroups have recently started to work on detect-and-avoid (DAA) mechanisms. According to [124] and [125] both subgroups 3a and 4a envision a frequency domain approach for DAA. Different from these approaches the temporal cognitive medium access, which we partially presented in [82] and [126], is proposed as interference avoidance mechanism for UWB WBANs. The temporal cognitive MAC can also be considered as a DAA in time-domain.

13.1 Principle Idea of the Temporal Cognitive MAC

As shown in Section 11, interference has a strong impact on UWB systems. However, it can be seen that a number of interferers transmit their data burst-wise. Hence, a UWB MAC can make use of this burst wise interference structure. In the following, the idea of such a MAC scheme is presented followed by a section with performance results.

It is assumed that the UWB device has a sleep mode. After a wake up the UWB device senses the channel with a kind of received signal strength indication (RSSI) and it directly transmits its data if the channel is not occupied by any interferer. Since strict latency time requirements have to be fulfilled for a number of applications, the UWB device has to transmit its data in a given latency time p. Each UWB packet has a packet duration b. Within each UWB packet (diagonal lined in Fig. 13.1) a preamble of duration x is present, e.g., it is used for synchronization issues. The number of unoccupied time slots N in a given latency time is determined by the interferers burst structure and the latency time, e.g., $N = 2$ for the example in Fig. 13.1. The duration of the ith unoccupied time slot is given by k_i. Since UWB pulse widths are below 2 ns, using an RSSI not only the UWB receiver but also the UWB transmitter can eavesdrop for any interferer in the time between two transmit pulses. Thus, a DAA mechanism in time domain can be established which avoids interference from UWB to existing wireless systems.

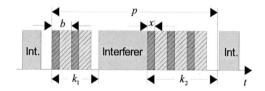

Figure 13.1: Scheme to determine the usable idle time with interfering bursts (solid) and UWB packets (lined)

The usable idle time, i.e., the time which can be used for UWB data transmission, are considered in the following for the evaluation of the TC MAC. The preamble and the time for UWB transmission in colliding packets are not considered as usable idle time. Based on the assumptions above the following expression for the average usable idle time $t_{\text{idle}}(b)$ can be achieved:

$$t_{\text{idle}}(b) = \frac{b-x}{p} \cdot \mathcal{E}\left\{\sum_{i=1}^{N} \left\lfloor \frac{k_i}{b} \right\rfloor\right\}, \tag{13.1}$$

where $\lfloor \cdot \rfloor$ rounds the argument to the next smaller integer. The expectation $\mathcal{E}\{\cdot\}$ is taken over the different channel realizations. To get rid of the given latency time, a division by p yields the normalized idle time per second. It can be seen that the packet length b has strong impact on $t_{\text{idle}}(b)$. On one hand, the time for the payload per packet $t_{\text{payload}} = b - x$ increases with b, on the other hand, the number of packets per empty slot time $\lfloor \frac{k_i}{b} \rfloor$ decreases with increasing packet length. This shows that there exists an optimum packet length that yields the maximum usable idle time.

Since (13.1) is discontinuous, it is not possible to determine the optimum UWB packet length by calculation of the derivative. Therefore, an approximation for the achievable usable idle time is derived. $\lfloor \cdot \rfloor$ can be written as

$$\left\lfloor \frac{k_i}{b} \right\rfloor = \frac{k_i}{b} - c, \tag{13.2}$$

with $c \in [0, 1)$. By setting $c = 0$ an upper bound for the usable idle time in (13.1) can be derived as

$$t_{\text{idle,max}}(b) = \frac{b-x}{p} \cdot \mathcal{E}\left\{\sum_{i=1}^{N} \left(\frac{k_i}{b}\right)\right\}. \tag{13.3}$$

This expression corresponds to an integer number of UWB packets with packet length b plus one packet with a reduced packet length $b' = c \cdot b$ in each empty slot of duration k_i, i.e., the whole time between interferer bursts is used for UWB transmission. Please note that the preamble of the variable packets also scales with its packet length b'. The corresponding packet placement is shown exemplarily in Fig. 13.2 a).

A lower bound for the usable idle time in (13.1) can be achieved by setting $c = 1$.

$$t_{\text{idle,min}}(b) = \frac{b-x}{p} \cdot \mathcal{E}\left\{ \sum_{i=1}^{N} \left(\frac{k_i}{b} - 1 \right) \right\} \tag{13.4}$$

In this case, the whole idle time less the time for one UWB packet is always used for UWB transmission. Corresponding to the example in Fig. 13.2 a), the packet placement for the lower bound is shown in Fig. 13.2 b). Same as for the upper bound, the preamble length of the variable packets scale here, too. Instead of considering the maximum and minimum value of c a uniform

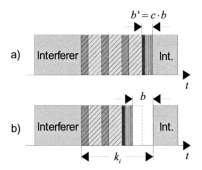

Figure 13.2: UWB packet distribution in an unoccupied slot of duration k_i, a) using $t_{\text{idle,max}}(b)$ from (13.3), and b) using $t_{\text{idle,min}}(b)$ from (13.4); Interferer bursts are plotted solid, UWB packets are plotted diagonal lined, and fractional UWB packets are vertical lined

distribution of c in $[0, 1)$ is assumed now, i.e., $\mathcal{E}\{c\} = \frac{1}{2}$. This means that half a UWB packet per idle time slot is lost on average. Using this expectation value, an approximation for the usable idle times in (13.1) is given by

$$t_{\text{idle,approx}}(b) = \frac{b-x}{p} \cdot \mathcal{E}\left\{ \sum_{i=1}^{N} \left(\frac{k_i}{b} - \frac{1}{2} \right) \right\}. \tag{13.5}$$

$t_{\text{idle,approx}}(b)$ is used to calculate the derivative and to determine the optimum UWB packet length. The optimum from (13.5) can be determined as

$$
\begin{aligned}
\frac{dt_{\text{idle,approx}}}{db} &= \frac{\bar{N}\mathcal{E}\{k_i\}x}{p\,b^2} - \frac{\bar{N}}{2p} = 0 \\
\Rightarrow \frac{\bar{N}}{2p} &= \frac{\bar{N}\mathcal{E}\{k_i\}x}{p\,b^2} \\
\Leftrightarrow b^2 &= \frac{2\bar{N}\mathcal{E}\{k_i\}x\,p}{\bar{N}p} = 2\mathcal{E}\{k_i\}x \\
\Rightarrow b &= \pm\sqrt{2E\{k_i\}x} = b_{\text{opt}}.
\end{aligned}
\tag{13.6}
$$

$\bar{N}$ denotes the mean value of N. In (13.6), only the positive solution for the UWB packet length makes sense. It can be seen that the preamble length x and the expectation over the length of idle time slots k_i determine the optimum UWB packet length. Although p is not explicitly contained in the expression for b_{opt} in (13.6), it is implicitly contained in $\mathcal{E}\{k_i\}$. Hence, the latency time p has an impact on the optimum packet length if p is smaller than the usual idle time between two interfering bursts.

13.2 Performance of the Temporal Cognitive MAC

In this section the performance of a UWB system is investigated with the temporal DAA as described in the previous section. We consider GSM, BT, and IEEE 802.11b WLAN as possible interferers.

13.2.1 Usable Idle Times and Packet Delays

Based on time domain measurements of these interference scenarios usable idle times for the UWB system are evaluated by using the expressions given in Section 13.1. It is assumed that each UWB packet contains a preamble of $x = 100\ \mu$s duration. For each interference scenario the measured usable idle time $t_{\text{idle}}(b)$ and the approximation $t_{\text{idle,approx}}(b)$ are shown. For completeness, the lower bound $t_{\text{idle,min}}(b)$ and the upper bound $t_{\text{idle,max}}(b)$ are also given. Since strict latency time limits have to be fulfilled in particular for speech transmission, moreover, the time is investigated that UWB packets are delayed if another wireless system is active. Therefore, the time delay between wake up of the UWB device and the beginning of the next successful transmitted UWB packet is shown for different packet lengths. We consider 50%, 10% and 1% outage-delay as well as the maximum occurred delay, which corresponds to a 0% outage. At the outage delays a certain percentage of all packets has a delay less than the given one. For example, the 1% outage-delay means that 99% of all UWB packets are transmitted successfully with a delay less than the determined one. All following evaluations are based on time domain measurements of the interference (see Fig. 10.3).

GSM In Fig. 13.3(a) and Fig. 13.3(b), the usable idle times are shown for a UWB system if it is interfered by one GSM device assuming latency times of $p = 2$ ms and $p = 5$ ms, respectively. The idle times are determined by applying the usable idle time equations given in Section 13.1 to 10000 randomly chosen measurement samples of duration p. It can be seen that $t_{\text{idle}}(b)$ lies for all UWB packet lengths between the lower bound $t_{\text{idle,min}}(b)$ and the upper bound $t_{\text{idle,max}}(b)$. The approximation $t_{\text{idle,approx}}(b)$ in Fig. 13.3(a) follows the trend of $t_{\text{idle}}(b)$. The steps and the zigzag behavior of $t_{\text{idle}}(b)$ are caused by the latency time $p = 2$ ms. Since GSM is transmitting a burst of 577 μs duration every 4.6 ms, the channel is idle during the latency time of 2ms for a number of measurement samples, i.e., in many cases 2 ms long UWB packets can be transmitted without collision. For packet lengths $b = \frac{p}{k}, k \in \mathbb{N}$, $t_{\text{idle,approx}}(b)$ exhibit steps. At these packet lengths a slight increase of the packet has the impact that one packet less can be transmitted in a given empty slot, which decreases the usable idle time significantly.

In Fig. 13.3(b), the usable idle times are shown for a latency time of $p = 5$ ms. There, the influence of the latency time is smaller, since the usable idle time is dominated by the GSM burst structure, which has an idle time of 4.1 ms. Thus, it can be seen that no UWB packets longer than this time can be transmitted. Up to the packet length of 4.1 ms $t_{\text{idle,approx}}(b)$ fits $t_{\text{idle}}(b)$ well. Since the maximum idle time is now longer than for $p = 2$ ms, the usable idle times and the optimum packet length are higher. For a latency time of $p = 2$ ms the maximum usable idle time of 583 ms/s is achieved with $b_{\text{opt,2ms}} = 544$ μs long UWB packets and for $p = 5$ ms the maximum idle time of 639ms/s is achieved with $b_{\text{opt,5ms}} = 693$ μs.

The evaluation of the time delays is also based on time domain measurements of the interference. 10000 wake up times of the UWB device are chosen randomly. For each wake

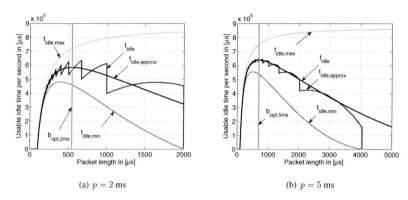

(a) $p = 2$ ms

(b) $p = 5$ ms

Figure 13.3: Usable idle times with one GSM interferer present, assuming $x = 100$ μs

up time the time delay between wake up and the beginning of the next successful transmitted UWB packet is determined. We do not consider the case that further UWB packets might be transmitted without any delay after the first successful transmission in an idle time slot. Such a consideration would result in lower time delays. Thus, the time delays shown in Fig. 13.4 for different packet lengths can be regarded as a worst case. It can be seen that there is a linear increase of the delays with increasing packet length. The maximum delay is given if a UWB packet slightly does not fit into the idle time slot before a GSM burst, i.e., the maximum delay is about $b + 577$ μs. This results in a linear increase of the time delays with increasing packet length. The packet delays for $b_{opt,5ms}$ are higher compared to $b_{opt,2ms}$. The maximum delays at $b_{opt,2ms}$ and $b_{opt,5ms}$ are about 1120 μs and 1269 μs, respectively. About 1% of all packets have a delay close to the maximum delay. However, due to the long idle times in GSM the 50% outage shows that 50% of all packets shorter than about 1800 μs are transmitted without any delay and longer packets are only slightly delayed.

BT In the range of short packet lengths, where the optimum packet length is located, $t_{idle}(b)$ and $t_{idle,approx}(b)$ match well as shown in Fig. 13.5(a). In the presence of a BT interferer the maximum usable idle time of 371 ms/s can be achieved by using packets of 372 μs duration. The usable idle time is mainly influenced by the burst structure, i.e., for packets longer than

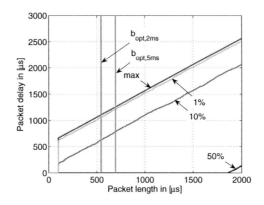

Figure 13.4: Packet delays with one GSM interferer present

884 μs the usable idle time drops to 0, because it is not possible to transmit any UWB packets. This limit is given by the maximum idle time of BT, which has a repetition time of 1250 μs and a burst length of 366 μs for a single-slot BT burst. During our measurements no multi-slot packets could have been observed. The zigzag behavior of the usable idle time is caused by the relatively short repetition of BT. For almost any position of the 2 ms latency time window there is one idle time slot with maximum idle time of 884 μs. Therefore, steps occur at fractions of this maximum idle time.

For the same reason as described above, the packet delays in Fig. 13.5(b) are only plotted up to a packet length of 884 μs. For shorter packet lengths the curves exhibit a linear increase, since the maximum delay is here given approximately by the maximum BT burst length plus the respective packet length, i.e., $b + 366$ μs. The maximum delay at b_{opt} is 737 μs. As it was the case for GSM 1% of all packets have a delay close to the maximum delay. Since BT bursts are shorter than GSM bursts, the maximum time delays are also shorter compared to GSM. However, the shorter repetition time of BT bursts leads to a larger number of UWB packets that are delayed. Thus, the 50% outage delay at b_{opt} is 105 μs. Only for UWB packets shorter than about 270 μs the 50% outage delay shows no delays.

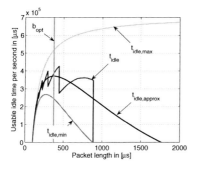

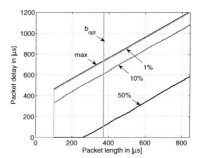

(a) Usable idle times with one BT interferer present, assuming $p = 2$ ms and $x = 100$ μs

(b) Packet delays with one BT interferer present

Figure 13.5: Consideration of BT interference

WLAN   The measured usable idle times for WLAN, which are shown in Fig. 13.6(a), do not exhibit such a zigzag behavior as the ones for GSM and BT. This is due to the non-periodic burst structure of WLAN. Due to this fact also the assumption $\mathcal{E}\{c\} = \frac{1}{2}$ holds and $t_{\text{idle}}(b)$ and $t_{\text{idle,approx}}(b)$ match very well. Nevertheless, the maximum usable idle time of 364ms/s, which can be achieved with packets of 407 μs duration, is close to the usable idle time that can be achieved if a BT device is interfering.

Due to the non-periodic burst structures of WLAN the packet delays are not anymore linear increasing with the packet length as it can be seen in Fig. 13.6(b). It can also happen that a packet is delayed for a relatively long time. By using UWB packets with the optimum packet length b_{opt} the maximum delay amounts to about 4697 μs. However, only a slight increase of the packet length yields a substantial increase of the maximum time delay. Since the measurement samples have a duration of only 10 ms, it can be observed that some packets with duration longer than about 550 μs cannot be transmitted with a delay of 10ms or less. For packet lengths of more than 929 μs and 1218 μs even 1% and 10% of all packets cannot be transmitted with a delay of 10 ms or less, respectively.

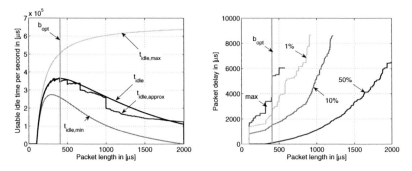

(a) Usable idle times with one WLAN interferer present, assuming $p = 2$ ms and $x = 100$ μs

(b) Packet delays with one WLAN interferer present

Figure 13.6: Consideration of WLAN interference

Conclusions   For the temporal cognitive MAC an expression for the usable idle time in the presence of burst interferers is presented. By using an approximation of this expression it is shown that the optimum UWB packet length is only determined by the preamble length, the expectation of the empty slot durations, and the latency time. It could have been observed that the optimum UWB packet length decreases with increasing occupied time of the channel. Although UWB systems using the temporal cognitive MAC can transmit only during a fraction of the whole time, it is shown that reasonable usable idle times and outage delays are achieved assuming GSM, BT, and WLAN as interferers. The usable idle times can be increased by introducing variable UWB packet lengths. As a result, theoretically the whole time between adjacent interferer bursts can be used for UWB transmission. However, such a usage of fully variable packet lengths would require a higher complexity than the ones of fixed packet lengths. For variable packet lengths the UWB devices require knowledge about the present interferers and their burst structures, i.e., their idle times and their burst durations. Complexity can be reduced by using only a set of possible packet lengths, which are chosen based on the channel observations in a certain time interval and the idle time statistics of the possible interferers. Besides the required information about the interferers the UWB devices have to exchange the data about their packet lengths, which can be included in the preamble, for instance. Since the usable idle times and packet delays are reasonable for UWB packets with fixed packet lengths and since variable packet lengths require higher complexity, fixed packet lengths are recommended for the use in low complexity UWB systems such as body area networks. However, variable packet lengths are interesting for systems where complexity is only secondary.

13.2.2 Packet Error Rates

In the previous sections the principle feasibility of the non-coherent receiver structures and the temporal cognitive medium access have been shown. However, no channel access has been specified, yet. Hence, in the following the packet error performance is investigated for different kinds of medium access schemes. For this evaluation a room of 100 m^2 size is assumed, in which N randomly placed UWB links are active. One exemplary distribution of $N = 10$ UWB links is shown in Fig. 13.7. The transmitter (TX) positions are marked by the $\times$ and the receiver (RX) positions by the $\circ$. The lines between the TXs and RXs indicate the TX-RX pairs that want to communicate. It is assumed that all TXs transmit with the same power. Between TX and RX a free space path loss according to equation (4.6) is considered. For evaluation, an interference limited scenario is assumed, i.e., additional noise is not considered. This path loss is used to calculate the signal-to-interference ratios at the receivers. A collision takes place if the SIR of a packet is below a given minimum SIR. For the evaluation of the packet error rates two cases are considered; a SIR of 0 dB and a SIR of 10 dB. This means that the interference power has to be smaller than the desired signal power for a required SIR of 0 dB and smaller than one tenth of the desired signal power for a SIR of 10 dB. If the SIR at the receiver is above the respective value, it is assumed that the impact of the interferer can be omitted by any kind of signal processing at the receiver.

1-persistent and non-persistent CSMA [127] are considered as medium access schemes with carrier sensing. A device using the 1-persistent CSMA scheme listens if the channel is occupied until it is idle and transmits its data [128]. In case of a collision a random backoff is performed. Using the non-persistent CSMA scheme, the device starts a random backoff if it detects an occupied channel or if a collision takes place. If the channel is idle, the device is directly

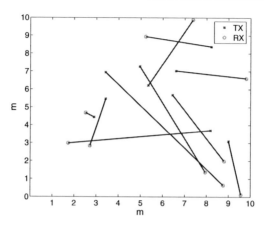

Figure 13.7: Exemplary distribution of $N = 10$ links within the considered room of size 10 m x 10 m

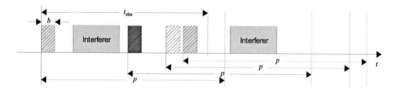

Figure 13.8: Exemplary UWB packet distribution for $N = 4$ with interferers

transmitting. For evaluation of the packer error rates collisions in the vulnerable period are also considered. Besides these both CSMA schemes ALOHA [129] is considered as well, because it might be an alternative for very simple transmit-only UWB nodes that do not have any channel sensing capabilities. Such nodes just wake-up and transmit their data.

For the evaluation of the packet error rates for N UWB links a setup is chosen as shown exemplarily for $N = 4$ links in Fig. 13.8. For the UWB packets a packet length $b = 303$ μs is chosen. The same duration, i.e., 303 μs, is chosen for both CSMA schemes as the maximum random backoff time. It is assumed that N UWB TX start to transmit their packet within the time interval $t_{\text{obs}} = 5$ ms. Each UWB packet has to be transmitted within a latency time $p = 5$ ms. In the case that a packet cannot be transmitted within this latency time it is considered as an erroneous packet.

In addition to the N UWB devices, GSM, BT, and IEEE 802.11b WLAN are considered as burst interferers that transmit simultaneously to the UWB devices. For each link a number of 10000 packets is transmitted. The PER curves are plotted over the number of UWB links for the different interference scenarios. The maximum number of UWB links is set to $N_{max} = 20$. Additionally, two cases are distinguished. In the first case, the impact of the interferers without bandpass filtering at the UWB RX is investigated, i.e., the interference power is so high that independent of the interferer and the UWB RX position a collision takes place. In the second case, a bandpass filter at the UWB RX is considered that attenuates the out-of-band interference. As described in section 10, it is assumed that only interferers within a sphere of 1 m distance are strong enough to cause a collision.

Packet Error Rates not considering Bandpass Filter   While packet errors for ALOHA have their origin only in collisions, packet errors for the CSMA schemes are only caused by hurting the latency time requirements. Using ALOHA, the PER is only below 10^{-1} if no interferer is present and only 2 UWB links are active. This can be seen from Fig. 13.9. In the presence of any interferer the PER curves are even for a small number of UWB links above 10^{-1}. The different required SIR values 0 dB and 10 dB have an impact on the no interference case only. Requiring a SIR of 10 dB, the PER is higher than for a SIR of 0 dB. The PERs in the presence of an interferer are unaffected by the different SIR values, since these PERs are dominated by the number of collisions with an interfering burst. The PER curves for WLAN and BT are almost the same. The PER performance of GSM is slightly better compared to these curves because the channel is less occupied by one GSM device than by a BT or a WLAN device. Hence, collisions with GSM are less likely than with BT or WLAN.

As it can be expected, both CSMA schemes exhibit a much better performance than the ALOHA scheme, because the UWB devices are able to eavesdrop the channel for present interferers and to resolve collisions. In Fig. 13.10, the performance of the 1-persistent CSMA scheme is depicted for both SIR values. For a required SIR of 0 dB no packet errors can be observed from Fig. 13.10(a) if no interferer is present, i.e., the PER is below $5 \cdot 10^{-4}$. In the

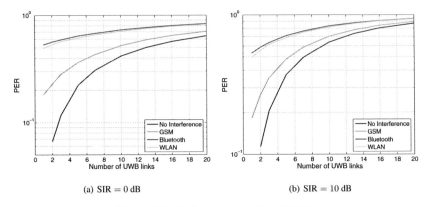

(a) SIR $= 0$ dB   (b) SIR $= 10$ dB

Figure 13.9: Packet error rate using ALOHA

presence of a GSM interferer the PER is below this value for less than 15 UWB links. However, with increasing number of nodes the PER increases up to almost 10^{-2} for 20 UWB links. Due to the higher duty cycles, i.e., smaller channel idle times of BT and WLAN, the PERs in the presence of such interferers are much worse. Assuming a minimum required PER of 10^{-2} up to 7 and 13 UWB links are possible in the presence of an active WLAN and BT device, respectively. The PER for WLAN is much worse than for BT since UWB packets may be delayed due to a WLAN interferer much more than due to a BT interferer as shown in Fig. 13.5(b) and Fig. 13.6(b). Considering a SIR of 10 dB the PER curves are shown in Fig. 13.10(b). There, it can be seen that the PER curves become worse. However, the number of UWB links for a minimum PER of 10^{-2} remains almost the same. In the case of no interference the PER is always below 10^{-2} for up to 20 UWB devices. If a GSM interferer is present, up to 18 UWB links are possible with a PER of less than 10^{-2}. While the number of UWB links in the presence of a BT devices reduces to 11 it stays 7 for a WLAN interferer. The PER for the WLAN case has almost the same performance as for the required SIR of 0 dB. This shows that this PER is mainly determined by the collisions with the interferer.

The PER curves for the non-persistent CSMA are worse compared to the 1-persistent CSMA as it can be seen in Fig. 13.11. This performance degradation is caused by a higher number of

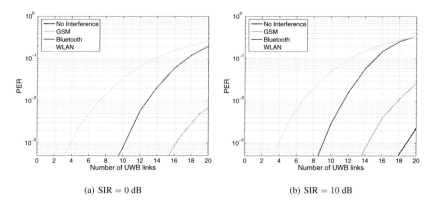

(a) SIR = 0 dB   (b) SIR = 10 dB

Figure 13.10: Packet error rates using 1-persistent CSMA

packets not meeting the latency time requirements. Since each node enters the random backoff mode in case of sensing an occupied channel, packets can be strongly delayed. Assuming a SIR of 0 dB the PER curves for the non-persistent CSMA are very similar to the ones for the 1-persistent CSMA, which assume a SIR of 10 dB. For the non interference case a PER of above 10^{-3} can be observed from Fig. 13.10(a) for more than 18 UWB links. However, the PER stays still below 10^{-2} for the maximum number of considered UWB links. In the presence of a GSM interferer up to 18 UWB links can be active with less than the required PER. Up to 6 and 11 UWB links are possible in the presence of a WLAN and BT interferer, respectively. If a SIR of 10 dB is required, these both PER curves become only slightly worse. This shows that these PER curves are mainly determined by collisions with the interferers and not by collisions between UWB packets. For the non-interference and the GSM case the picture looks different. There, the PER curves get worse if a higher SIR is required. However, still 14 and 17 UWB links can be active while fulfilling the PER requirement of 10^{-2}.

Since the packet error rates using ALOHA are too high for a reliable transmission, it is inevitable to use one of the presented CSMA schemes together with the temporal cognitive MAC, if no bandpass filter is used to attenuate the out-of-band interferers. In the case of the existence of a high number of UWB links the 1-persistent CSMA should be used due to its much

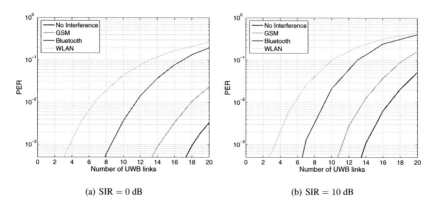

(a) SIR = 0 dB   (b) SIR = 10 dB

Figure 13.11: Packet error rates using non-persistent CSMA

better performance rather than the non-persistent CSMA. However, for a moderate number of links both CSMA schemes perform very well and they exhibit packet errors rates below 10^{-2}.

Packet Error Rates considering Bandpass Filter From the previous investigations in some cases it could have been observed that the packet error rates are determined mainly by the out-of-band interferer. Hence, a bandpass filter as proposed in section 9 is considered in the following. By using such a filter a UWB packet is only disturbed, if the interferer is less than 1 m away from the UWB receiver. Hence, for the following evaluations of the PER curves also the distances between the interferer and the UWB receivers are considered. Although the PER curves for the non-interference case do not change, they are presented in the following for comparison, too.

For ALOHA the PER curves in the presence of any interferer approach the PER curve without interferer, which is shown in Fig. 13.12. Assuming the bandpass filter, packet errors are mainly caused by collisions between UWB packets and not by collisions with the interferers anymore. Anyway, the packet error rates are above 10^{-1} for more than 2 UWB links.

The PER curves for 1-persistent CSMA considering a bandpass filter in Fig. 13.13 are much better than the ones without filter in Fig. 13.10. In the considered range of up to 20 UWB links, the PER curves are always below 10^{-2} for a SIR of 0 dB except for the WLAN case. There,

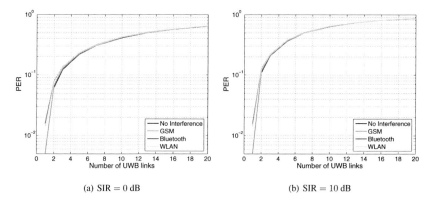

(a) SIR = 0 dB   (b) SIR = 10 dB

Figure 13.12: Packet error rate using ALOHA using a bandpass filter to reduce out-of-band interference

the PER is below the required one for less than 17 UWB links. Requiring a SIR of 10 dB at the UWB receivers the PER curves become worse compared to the less demanding SIR constraint. However, the PER is below the required value for less than 15 and 19 UWB links in the presence of a WLAN and BT interferer, respectively. In the presence of a GSM device or no interferer the PER is below 10^{-2} for the maximum number of considered UWB links and the PER in the presence of GSM is only slightly worse compared to the no interference case.

Using non-persistent CSMA the PER curves in the presence of a GSM interferer are almost the same as the ones without interference. This can be observed for both SIR values in Fig. 13.14. The PER curves in the presence of a BT interferer are also only slightly worse. Only if a WLAN interferer is present, the impact on the PER is higher. However, even for 20 UWB links in the presence of a WLAN, the PER is about 10^{-2}. For all other scenarios, the PER is below this value. Requiring a SIR of 10 dB a PER of less than 10^{-2} can be seen for less than 15 - 17 UWB links depending on the interferer present. These values are much higher compared to the ones if no bandpass filter is considered at the UWB receiver.

As in the case where no bandpass filter is considered at the UWB receiver, the performance of the ALOHA scheme is not sufficient for a large number of UWB links. It might be only

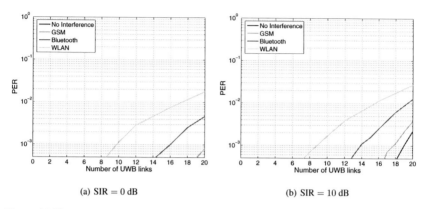

(a) SIR = 0 dB   (b) SIR = 10 dB

Figure 13.13: Packet error rates using 1-persistent CSMA using a bandpass filter to reduce out-of-band interference

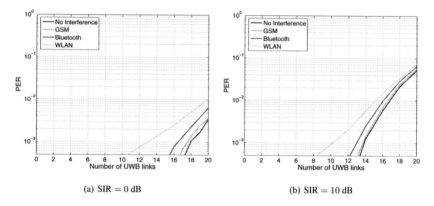

(a) SIR = 0 dB   (b) SIR = 10 dB

Figure 13.14: Packet error rates using non-persistent CSMA using a bandpass filter to reduce out-of-band interference

suited for a single link or maybe 2 links if the packet error rate requirements can be relaxed. For the CSMA schemes the number of UWB links assuming a maximum PER of 10^{-2} can be substantially increased by using a bandpass filter at the receiver to reduce out-of-band interference.

Part VI

Conclusions

Chapter 14

Recommendations

The body area network channel as well as potential interferers to UWB have been investigated in the previous sections. Moreover, a family of receiver structures has been presented for binary pulse position modulation, transmitted reference pulse amplitude modulation, and transmitted reference pulse interval amplitude modulation. These receiver structures have been derived by finding the respective maximum likelihood receiver assuming a different level of channel state information. Finally, the temporal cognitive medium access control has been introduced and the performance evaluated.

Based on these results, recommendations for a BAN system model are presented in the following.

14.1 Antenna

From the channel measurements it has been observed that signals radiated into the body are severely attenuated. Therefore, the antenna should not radiate into the direction of the body. Since the positions of transmitters and receivers in body area networks are generally not known, the antenna pattern shall have an omnidirectional characteristics. However, if there is no communication foreseen with other off-body devices in the vicinity of the body area network, the antenna could be designed such that the energy is radiated only along the body surface and neither into the body or away from it. With such an antenna also the interference between different body area networks as well as the interference from other wireless devices in the vicinity of the body area network could be reduced.

14.2 Receiver Structure

For a wireless body area network a receiver structure should not only be very simple, but also have a good performance. It has been shown that there is a tradeoff between performance and complexity, which is required for the acquisition of the channel state information as well as for the more complex receiver structure itself. Since the receiver structures for binary PPM and TR PAM have the same bit error performance, the binary PPM is suited better for the use in wireless body area networks due to the simpler receiver realizations. Of course, the energy detector, i.e., the optimum receiver for binary PPM without CSI knowledge, is the simplest possible receiver realization. For the use in BANs the integration duration should be very short in such receivers. However, the performance is only suboptimal in the case of an inappropriate chosen integration duration or in the presence of an interferer. Hence, the energy detector is only recommended in the case that the complexity is the most stringent requirement. Due to the only moderate complexity increase, which is required for the estimation of the average power delay profile, the binary PPM receiver with APDP knowledge yields a good compromise between performance and complexity. Moreover, this receiver is more robust against inter-symbol interference and an inappropriately chosen integration duration, since the weighting can be regarded as an automatic adaptation of the integration duration. Receivers with higher CSI level cannot be recommended due to the complex channel estimation that is required. This is also the case for the TR PIAM receiver structures. In the case that the bit error performance is the most stringent requirement the TR PIAM receiver with APDP knowledge is recommended. The performance is better than the one of the binary PPM receiver with APDP knowledge. However, this performance improvement requires a higher design complexity.

14.3 Interference Mitigation

Since the maximum transmit power allowed for UWB is very small, UWB devices have to cope with interference from other wireless services that have a much higher transmit power. Therefore, out-of-band interferers can also be harmful for UWB communications. It has been shown in the previous sections that the temporal cognitive medium access control is generally well suited to avoid interference from burst wise transmitting systems. In combination with the ALOHA scheme the packet error rates are too high for reliable communication. Although the 1-persistent CSMA scheme shows the best packet error rate performance, the non-persistent CSMA scheme is recommended in conjunction with the temporal cognitive MAC, since the energy efficiency is better. A device using the 1-persistent CSMA scheme has to eavesdrop the channel during the whole time after determining that it is occupied while the non-persistent CSMA scheme can turn into sleep mode during the random backoff time. However, this temporal cognitive MAC is not sufficient for continuously transmitting devices such as UMTS in the FDD mode. Another mitigation technique is required to avoid this kind of interferers. It is recommended to avoid UMTS interferers by introducing a notch filter that suppresses the corresponding frequency range. This could also be achieved by choosing an appropriate design of the antennas. Besides this, the antennas should have a bandpass characteristics, which allows for attenuation of out-of-band interferers.

Chapter 15

Summary and Conclusion

Based on 1100 channel measurements in the frequency range from 2 to 8 GHz a body area network channel model has been derived. It has been shown by using the Akaike information criterion that the magnitudes of the channel impulse responses are log-normal distributed. From theory, measurements, and simulations it has been observed that no direct transmission through the body takes place. Hence, antennas should be designed such that they do not radiate into the body and rather away from it or along its surface. Moreover, it has been shown that the channel is robust against distance variations between the antennas and the skin relaxing the constraints on the antenna mounting at the body. However, for the ear-to-ear link different antenna positions have been considered identifying the behind ear position as the one with the smallest attenuation. For communication around the human body the most energy is contained in a short window at the beginning of the channel impulse responses. This indicates that a very short integration time in a non-coherent receiver is sufficient to collect a large amount of the energy available in the channel.

Since UWB systems only have a very small transmit power and since the path loss for transmission at the human body might be high, interference from other wireless systems is an important issue. For investigation purposes the interference has been distinguished in two classes. Interferers which are active almost all the time have been referred to as background noise while interferers which only transmit periodically have been referred to as burst noise. It has been shown by measurements that GSM and UMTS base stations are the most severe background interferers. While this type of interference can be mitigated by using a filter, another strategy has been found for burst interferers, since these interferers can be in close vicinity of a UWB device. To avoid the burst interference it has been proposed to use the time between adjacent interferer bursts for UWB transmission. This scheme has been referred to as temporal cognitive medium access. Based on time domain measurements of the interferer bursts it has been shown that reasonable usable idle times can be achieved. The UWB device can use this time for transmission with the temporal cognitive medium access. Moreover, an expression for the optimum UWB packet length has been presented yielding the maximum data rate. Considering different combinations of bandpass filter, access scheme, and temporal cognitive MAC packet error rates have been evaluated. It could be observed that packet error rates below 10^{-2} can be obtained for up to about 15 parallel UWB links if a bandpass filter is used for mitigation of out-of-band interference.

For a UWB body area network the hardware should be as simple as possible. Therefore, non-coherent receiver structures such as an energy detector and a transmitted reference receiver are very promising solutions for the use in body area networks. Although the modulation alphabets that can be used are different, both receiver structures are very similar. It has been observed that the integration time has a strong impact on the performance. Since both receiver structures lose some dB of performance when compared with a coherent receiver structure, maximum likelihood receivers for binary pulse position modulation, transmitted reference pulse amplitude modulation, and transmitted reference pulse interval amplitude modulation have been derived assuming different amount of channel state information. By these means, it has been shown that the well known transmitted reference receiver and the energy detector are optimal if no channel state information is available. Evaluating the performance of the receiver structures with different channel state information level a trade-off between complexity and performance has been observed. The receiver structure with knowledge of the average power delay profile for binary pulse position modulation has been identified as the best suited solution for body area networks. For most channels this receiver yields better performance than the one without channel state information. Moreover, the receiver with average power delay profile knowledge is robust against an inappropriately chosen integration duration. These advantages are achieved at the cost of a moderate complexity increase.

Further research activities could concern the investigation of simple synchronization schemes. There is a number of solutions for synchronization but often they do not consider the BAN specific properties and requirements or specific receiver structures.

Most measurements in this thesis have been performed in the anechoic chamber. Even though some measurements have been done in an office room, where the same aggregation of energy could be observed as in the anechoic chamber, further measurements in different environment could be performed. These measurements would provide also insights, which data rates can be achieved and how robust receivers should be against inter-symbol interference.

For the temporal cognitive MAC only fixed packet lengths have been considered. Since the typical interferers for UWB are known, statistical information about the interference could be included in the temporal cognitive MAC. This would probably lead to UWB packets with variable packet lengths.

A Distribution Functions

A.1 Rayleigh Distribution

For $x \geq 0$, the Rayleigh distribution [111] is defined as

$$f(x) = \frac{x}{\sigma^2} e^{-\frac{x^2}{2\sigma^2}}. \tag{A.1}$$

A.2 Nakagami Distribution

The Nakagami distribution [111] is defined for $x > 0$ as

$$f(x) = \frac{2}{\Gamma(m)} \left(\frac{m}{\Omega}\right)^m x^{2m-1} e^{-\frac{m}{\Omega}x^2}. \tag{A.2}$$

where $m \geq 0.5$ is the Nakagami m-factor, $\Gamma(m)$ denotes the gamma function, and Ω denotes the mean-square value of the amplitude. For $m = 1$, the Nakagami distribution equals the Rayleigh distribution.

A.3 Weibull Distribution

According to [111], the Weibull distribution is defined for $x \geq 0$ as

$$f(x) = \alpha x^{\beta-1} e^{-\frac{\alpha}{\beta}x^\beta}. \tag{A.3}$$

The Rayleigh distribution is a special case of the Weibull distribution if $\alpha = \frac{1}{\sigma^2}$ and $\beta = 2$.

A.4 Rice Distribution

The Rice distribution [130] is defined as

$$f(x) = \frac{2(K+1)x}{\Omega} e^{-K - \frac{(K+1)x^2}{\Omega}} \cdot I_0 \left(2x \sqrt{\frac{K(K+1)}{\Omega}} \right) \tag{A.4}$$

where K denotes the Ricean K factor and I_0 denotes the 0th-order modified Bessel function of the first kind. If $K = 0$, the Rice distribution reduces to a Rayleigh distribution.

A.5 Lognormal Distribution

The lognormal distribution [111] is defined as

$$f(x) = \frac{1}{\sigma x \sqrt{2\pi}} e^{-\frac{(\ln(x) - \mu)^2}{2\sigma^2}}. \tag{A.5}$$

Nomenclature

$\cosh(x)$	Hyperbolic cosine
γ	Path loss exponent
$\Im\{x\}$	Imaginary part of x
λ	Wavelength
$\lceil \cdot \rceil$	Round up to the next higher integer number
$\lfloor \cdot \rfloor$	Rounding to the next smaller integer
$\mathcal{E}\{x\}$	Expectation of x
μ_0	Permeability in free space
μ_r	Relative permeability
ω	Angular frequency
$\Re\{x\}$	Real part of x
mod	Modulo operator
ε_0	Permittivity in free space
ε_r	Relative permittivity

B	Bandwidth
BW	Fractional bandwidth
C	Capacity
c	Light speed
E_p	Peak energy
E_s	Signal energy
E_{sp}	Energy density of a single pulse
f_c	Center frequency
f_l	Lower cut-off frequency
f_u	Upper cut-off frequency
$g(t)$	Filter impulse response
$g(t) * s(t)$	Convolution of $g(t)$ and $s(t)$
$H(f)$	Frequency transfer function of the channel
$h(t)$	Channel impulse response
$n(t)$	Additive white Gaussian noise
$N_0/2$	Noise power spectral density
$p(t)$	UWB Pulse
P_{RX}	Receive power
P_{TX}	Transmit power
$r(t)$	Receive signal

$s(t)$	Transmit signal
$S^\star(f)$	Conjugate complex of $S(f)$
t	Time
$U(\theta)$	Energy pattern
ADC	Analog-to-digital converter
AGC	Adaptive gain control
AIC	Akaike information criterion
APDP	Average power delay profile
ARAKE	All RAKE
AWGN	Additive white gaussian noise
BAN	Body area network
BER	Bit error ratio
BT	Bluetooth
CCI	Co-channel interference
cdf	Cumulative distribution function
CIR	Channel impulse response
CMOS	Complementary metal oxide semiconductor
CSI	Channel state information
CSI	Channel state information
CSMA	Carrier sense multiple access

DAA	Detect-and-avoid
DARPA	Defense Advanced Research Projects Agency
DECT	Digital enhanced cordless telecommunications
DS	Direct sequence
ECC	Electronic Communications Committee
ED	Energy detector
EIRP	Equivalent isotropically radiated power
FCC	Federal Communications Commission
FDD	Frequency division duplex
FDMA	Frequency division multiple access
FDTD	Finite Difference Time Domain
GHz	Gigahertz
GPRS	General packet radio services
GRP	Glass-fiber reinforced plastic
GSM	Global system for mobile communications
HPF	High pass filter
IEEE	Institute of Electrical and Electronics Engineers
IPDP	Instantaneous power delay profile
IR	Impulse radio
ISI	Intersymbol interference

ISM bands	Industrial, scientific, and medical bands
ITU	International Telecommunication Union
kHz	Kilohertz
KL distance	Kullback-Leibler distance
LNA	Low noise amplifier
LOS	Line-of-sight
LTI	Linear time-invariant
MAC	Medium access control
MAP	Maximum-a-posteriori
MB-OFDM	multi-band orthogonal frequency division multiplexing
Mcps	Mega chips per second
MF	Matched filter
MHz	Megahertz
ML	Maximum likelihood
NBI	Narrowband interferer
NLOS	Non-line-of-sight
OFDM	Orthogonal frequency division multiplexing
OOK	On-off keying
PAM	Pulse amplitude modulation
PDA	Personal digital assistant

PDP	Power delay profile
PER	Packet error rate
PIAM	Pulse interval amplitude modulation
PL	Path loss
PLL	Phased-locked-loop
PPM	Pulse position modulation
PRAKE	Partial RAKE
PSD	Power spectral density
RF	Radio frequency
RSSI	Received signal strength indication
SAM	Specific Anthropomorphic Mannequin
SDR	Step recovery diode
SIR	Signal-to-interference ratio
SNR	Signal-to-noise ratio
SRAKE	Selective RAKE
TDD	Time division duplex
TDMA	Time division multiple access
TFC	Time-Frequency Coding
TH	Time-Hopping
TR	Transmitted reference

UE	User equipment
UMTS	Universal Mobile Telecommunications System
UWB	Ultra Wideband
VCO	Voltage controlled oscillator
VNA	Vector network analyzer
WBAN	Wireless body area network
WCDMA	Wideband code division multiple access
WLAN	Wireless local area network
WPAN	Wireless personal area network

Bibliography

[1] Raymond Knopp and Younes Souilmi. Achievable rates for UWB peer-to-peer networks. *International Zurich Seminar on Communications, IZS*, pages 82–85, February 2004.

[2] Liuqing Yang. Rate-scalable UWB for WPAN with heterogeneous nodes. *IEEE International Conference on Acoustics, Speech, and Signal Processing, ICASSP*, 3:625–628, March 2005.

[3] David Barras, Frank Ellinger, and Heinz Jäckel. Comparison between ultra-wideband and narrowband transceivers. *TRLabs/IEEE Wireless*, July 2002.

[4] Liuqing Yang and Georgios B. Giannakis. Ultra-wideband communications: an idea whose time has come. *IEEE Signal Processing Magazine*, 21(6):26–54, November 2004.

[5] Gopal Racherla, Jason L. Ellis, David S. Furuno, and Susan C. Lin. Ultra-wideband systems for data communications. *IEEE International Conference on Personal Wireless Communications*, pages 129–133, December 2002.

[6] Thad. B. Welch, Randall L. Musselman, Bomono A. Emessiene, Phippip D. Gift, Daniel K. Choudhury, Derek N. Cassadine, and Scott M. Yano. The effects of the human body on UWB signal propagation in an indoor environment. *IEEE Journal on Selected Areas in Communications*, 20(9):1778–1782, December 2002.

[7] Thomas Zasowski, Frank Althaus, Mathias Stäger, Armin Wittneben, and Gerhard Tröster. UWB for noninvasive wireless body area networks: Channel measurements

and results. *IEEE Conference on Ultra Wideband Systems and Technologies, UWBST*, pages 285–289, November 2003.

[8] István Z. Kovács, Gert F. Pedersen, Patrick C. F. Eggers, and Kim Olesen. Ultra wideband radio propagation in body area network scenarios. *IEEE Eighth International Symposium on Spread Spectrum Techniques and Applications, ISSSTA*, pages 102–106, September 2004.

[9] Andrew Fort, Claude Desset, Julien Ryckaert, Philippe De Doncker, Leo Van Biesen, and Stephane Donnay. Ultra wide-band body area channel model. *IEEE International Conference on Communications, ICC*, 4:2840–2844, May 2005.

[10] Andrew Fort, Claude Desset, Julien Ryckaert, Philippe De Doncker, Leo Van Biesen, and Piet Wambacq. Characterization of the ultra wideband body area propagation channel. *IEEE International Conference on Ultra-Wideband, ICU 2005*, pages 22–27, September 2005.

[11] Andrew Fort, Claude Desset, Philippe De Doncker, Piet Wambacq, and Leo Van Biesen. An ultra-wideband body area propagation channel model- from statistics to implementation. *IEEE Transactions on Microwave Theory and Techniques*, 54(4):1820–1826, April 2006.

[12] A. Alomainy, Y. Hao, C.G. Parini, and P.S. Hall. On-body propagation channel characterisation for UWB wireless body-centric networks. *IEEE Antennas and Propagation Society International Symposium*, 1b:694–697, July 2005.

[13] Steve Stroh. Ultra wideband: Multimedia unplugged. *IEEE Spectrum*, 40(9):23–27, September 2003.

[14] Ian Oppermann. The role of UWB in 4G. *Wireless Personal Communications*, 29(1-2):121–133, April 2004.

[15] Craig K. Rushforth. Transmitted-reference techniques for random or unknown channels. *IEEE Transactions on Information Theory*, 10(1):39–42, January 1964.

[16] Harry Urkowitz. Energy detection of unknown deterministic signals. *Proceedings of the IEEE*, 55(4):523–531, April 1967.

[17] Ralph Hoctor and Harold Tomlinson. Delay-hopped transmitted-reference RF communications. *IEEE Conference on Ultra Wide Band Systems and Technologies, UWBST*, pages 265–269, May 2002.

[18] Yi-Ling Chao and Robert A. Scholtz. Optimal and suboptimal receivers for ultra-wideband transmitted reference systems. *IEEE Global Telecommunications Conference, Globecom*, 2:759–763, December 2003.

[19] Farid Dowla, Faranak Nekoogar, and Alex Spiridon. Interference mitigation in transmitted-reference ultra-wideband (UWB) receivers. *IEEE Antennas and Propagation Society International Symposium*, 2:1307–1310, June 2004.

[20] Antonio A. D'Amico and Umberto Mengali. GLRT receivers for UWB systems. *IEEE Communications Letters*, 9(6):487–489, June 2005.

[21] Jin Tang and Zhengyuan Xu. A novel modulation diversity assisted ultra wideband communication system. *IEEE International Conference on Acoustics, Speech, and Signal Processing, ICASSP*, 3:309–312, March 2005.

[22] Thomas Zasowski, Frank Althaus, and Armin Wittneben. An energy efficient transmitted-reference scheme for ultra wideband communications. *International Workshop on Ultra Wideband Systems joint with Conference on Ultrawideband Systems and Technologies, Joint UWBST & IWUWBS*, pages 146–150, May 2004.

[23] Geert Leus and Alle-Jan van der Veen. Noise suppression in UWB transmitted reference systems. *IEEE 5th Workshop on Signal Processing Advances in Wireless Communications*, pages 155–159, July 2004.

[24] Jac Romme and Guiseppe Durisi. Transmit reference impulse radio systems using weighted correlation. *International Workshop on Ultra Wideband Systems joint with*

Conference on Ultrawideband Systems and Technologies, Joint UWBST & IWUWBS, pages 141–145, May 2004.

[25] Jac Romme and Klaus Witrisal. Transmitted-reference UWB systems using weighted autocorrelation receivers. *IEEE Transactions on Microwave Theory and Techniques*, 54(4):1754–1761, April 2006.

[26] Tony Q. S. Quek and Mow Z. Win. Ultrawide bandwidth transmitted-reference signaling. *IEEE International Conference on Communications, ICC*, 6:3409–3413, 20-24 June 2004.

[27] Tony Q. S. Quek and Moe Z. Win. Analysis of UWB transmitted-reference communication systems in dense multipath channels. *IEEE Journal on Selected Areas in Communications*, 23(9):1863–1874, September 2005.

[28] Yi-Ling Chao and Robert A. Scholtz. Ultra-wideband transmitted reference systems. *IEEE Transactions on Vehicular Technology*, 54(5):1556–1569, September 2005.

[29] Emil Jovanov, Amanda O'Donnell Lords, Dejan Raskovic, Paul G. Cox, Reza Adhami, and Frank Andrasik. Stress monitoring using a distributed wireless intelligent sensor system. *IEEE Engineering in Medicine and Biology Magazine*, 22(3):49–55, May-June 2003.

[30] Bert Gyselinckx, Chris Van Hoof, Julien Ryckaert, Refet Firat Yazicioglu, Paolo Fiorini, and Vladimir Leonov. Human++: Autonomous wireless sensors for body area networks. *IEEE 2005 Custom Integrated Circuits Conference*, pages 13–19, September 2005.

[31] Holger Junker, Mathias Stäger, Gerhard Tröster, Dmitri Blättler, and Olivier Salama. Wireless networks in context aware wearable systems. *1st European Workshop on Wireless Sensor Networks*, pages 37–40, January 2004.

[32] Johann Chiang, Waqas Akram, Trent Johnson, Xu Gao, and Rajiv Bhatia. A wireless motion system for video gaming. *International Conference on Consumer Electronics*, pages 425–426, January 2005.

[33] Jianchu Yao, Ryan Schmitz, and Steve Warren. A wearable point-of-care system for home use that incorporates plug-and-play and wireless standards. *IEEE Transactions on Information Technology in Biomedicine*, 9(3):363–371, September 2005.

[34] Florian Trösch and Thomas Zasowski. The most important facts you have to know about UWB: An introduction. *Internal Report*, September 2006.

[35] Dieter J. Cichon and Werner Wiesbeck. The Heinrich Hertz Experiments at Karlsruhe in the View of Modern Communication. *International Conference on 100 Years of Radio*, pages 1–6, September 1995.

[36] Mohammad Ghavami, Lachlan B. Michael, and Ryuji Kohno. *Ultra wideband signals and systems in communication engineering*. Wiley, 1st edition, 2004.

[37] Office of the Secretary of Defense, Defense Advanced Research Projects Agency (DARPA). Assessment of Ultra-Wideband (UWB) Technology. *R-6280 (1990), OSD/DARPA, Ultra-wideband Radar Review Panel*, 1990.

[38] Manfred Thumm. Historical german contributions to physics and applications of electromagnetic oscillations and waves. *International Conference on Progress in Nonlinear Science*, pages 1–6, July 2001.

[39] FCC. Revision of part 15 of the commission's rules regarding ultra-wideband transmission systems. *First Report and Order, ET Docket 98-153, FCC 02-48*, adopted/released Feb. 14/ Apr. 22 2002.

[40] Electronic Communications Committee. ECC Decision of 24 March 2006 on the harmonised conditions for devices using Ultra-Wideband (UWB) technology in bands below 10.6 GHz. *(ECC/DEC/(06)04)*, March 2006.

[41] Commission on the European Communities. Commission Decision of 21/II/2007 on allowing the use of the radio spectrum for equipment using ultra-wideband technology in a harmonised manner in the Community. February 2007.

[42] Tan Geok Leng. Singapore UWB programme framework. *Internal report*, March 2003.

[43] Ryuji Kohno. Interpretation and future modification of Japanese regulation for UWB. *IEEE 802.15-06-0140-01-004a*, March 2006.

[44] Domenico Porcino and Walter Hirt. Ultra-wideband radio technology: potential and challenges ahead. *IEEE Communications Magazine*, 41(7):66–74, July 2003.

[45] Claude E. Shannon. A mathematical theory of communication. *The Bell Systems Technical Journal*, 27:379–423,623–656, July, October 1948.

[46] Harald T. Friis. A note on a simple transmission formula. *Proceedings of the I.R.E and Waves and Electrons*, pages 254–256, May 1946.

[47] Sinan Gezici, Zhi Tian, Georgios B. Giannakis, Hisashi Kobayashi, Andreas F. Molisch, H. Vincent Poor, and Zafer Sahinoglu. Localization via ultra-wideband radios: a look at positioning aspects for future sensor networks. *IEEE Signal Processing Magazin*, pages 70–84, July 2005.

[48] Moe Z. Win and Robert A. Scholtz. Impulse radio: how it works. *IEEE Communications Letters*, 2(2):36–38, February 1998.

[49] Jaiganesh Balakrishnan, Anuj Batra, and Anand Dabak. A multi-band OFDM system for UWB communication. *IEEE Conference on Ultra Wideband Systems and Technologies*, pages 354–358, November 2003.

[50] Jeff R. Foerster et. al. Channel modeling sub-comitee report final. *IEEE P802.15 WG for WPANs Technical Report, no. 02/490r0-SG3a*, 2002.

[51] Andreas F. Molisch, Jeffrey R. Foerster, and Marcus Pendergrass. Channel models for ultrawideband personal area networks. *IEEE Transactions on Wireless Communications*, 10(6):14–21, December 2003.

[52] Adel A. M. Saleh and Reinaldo A. Valenzuela. A statistical model for indoor multipath channels. *IEEE Journal on Selected Areas in Communications*, 5:128–137, February 1987.

[53] Andreas F. Molisch, Kannan Balakrishnan, Chia-Chin Chong, Shahriar Emami, Andrew Fort, Johan Karedal, Jürgen Kunisch, Hans Schantz, Ulrich Schuster, and Kai Siwiak. Ieee 802.15.4a channel model - final report. Technical report, IEEE P802.15 02/490r1–SG3a, September 2004.

[54] John G. Proakis. *Digital Communications*. McGraw-Hill Higher Education, 4th edition, 2001.

[55] R. Price and P. E. Green. A communication technique for multipath channels. *Proceedings of the IRE*, 46:555–570, March 1958.

[56] Dajana Cassioli, Moe Z. Win, Francesco Vatalaro, and Andreas F. Molisch. Performance of low-complexity rake reception in a realistic UWB channel. *IEEE International Conference on Communications 2002, ICC 2002*, 2:763–767, April-May 2002.

[57] Matt Welborn. TG4a general framework. *IEEE 802.15-05/0039r0*, January 2006.

[58] Robert A. Scholtz. Multiple access with time-hopping impulse modulation. *IEEE Military Communications Conference, MILCOM'93*, 2:447–450, October 1993.

[59] Jeongwoo Han and Cam Nguyen. A new ultra-wideband, ultra-short monocycle pulse generator with reduced ringing. *IEEE Microwave and Wireless Components Letters*, 12(6):206–208, June 2002.

[60] Lucian Stoica, Sakari Tiuraniemi, Ian Oppermann, and Heikki Repo. An ultra wideband low complexity circuit transceiver architecture for sensor networks. *IEEE International Symposium on Circuits and Systems, ISCAS*, 1:364–367, May 2005.

[61] Takahide Terada, Shingo Yoshizumi, Muhammad Muqsith adn Yukitoshi Sanada, and Tadahiro Kuroda. A CMOS ultra-wideband impulse radio transceiver for 1-Mb/s data

communications and ±2.5-cm range finding. *IEEE Journal of Solid-State Circuits*, 41(4):891–898, April 2006.

[62] Shingo Yoshizumi, Takhide Terada, Jun Furukawa, Yukitoshi Sanada, and Tadahiro Kuroda. All digital transmitter scheme and transceiver design for pulse-based ultra-wideband radio. *IEEE Conference on Ultra Wideband Systems and Technologies, UWBST*, pages 438–442, November 2003.

[63] Moe Z. Win and Robert A. Scholtz. Energy capture vs. correlator resources in ultra-wide bandwidth indoor wireless commuications channels. *IEEE Military Communications Conference, MILCOM'97*, 3:1277–1281, November 1997.

[64] Martin Weisenhorn and Walter Hirt. Robust noncoherent receiver exploiting UWB channel properties. *International Workshop on Ultra Wideband Systems joint with Conference on Ultrawideband Systems and Technologies, Joint UWBST & IWUWBS*, pages 156–160, May 2004.

[65] Lucian Stoica, Alberto Rabbachin, Heikki Olavi Repo, Teemu Sakari Tiuraniemi, and Ian Oppermann. An ultra wideband system architecture for tag based wireless sensor networks. *IEEE Transactions on Vehicular Technology*, 54(5):1632–1645, September 2005.

[66] Skycross SMT-3TO10M, 3.1-10 GHz Ultra-Wideband Antenna. [Online]. Available: http://www.skycross.com/Products/PDFs/SMT-3TO10M-A.pdf, March 05, 2007.

[67] Schmid & Partner Engineering AG, SPEAG, Zurich, Switzerland. [Online]. Available: http://www.speag.com.

[68] David L. Means and Kwok W. Chan. Evaluating compliance with FCC guidelines for human exposure to radiofrequency electromagentic fields, additional information for evaluating compliance of mobile and portable devices with FCC limits for human exposure to radiofrequency emissions. *FCC Supplement C (Edition 01-01) to OET Bulletin 65 (Edition 97-01)*, pages 1–53, June 2001.

[69] Andrew Fort, Julien Ryckaert, Claude Desset, Philippe De Doncker, Piet Wambacq, and Leo Van Biesen. Ultra-wideband channel model for communication around the human body. *IEEE Journal on Selected Areas in Communications*, 24(4):927–933, April 2006.

[70] Hirotugu Akaike. Information theory and an extension of the maximum likelihood principle. *Proceedings of the International Symposium on Information Theory*, pages 267–281, 1973.

[71] Kenneth P. Burnham and David R. Anderson. *Model Selection and Multimodel Interference: A Practical Informaion-Theoretic Approach*. Springer-Verlag, 2nd edition, 2002.

[72] Hirotugu Akaike. A new look at the statistical model identification. *IEEE Transactions on Automatic Control*, 19(6):716–723, December 1974.

[73] Hirotugu Akaike. On the likelihood of a time series model. *The Statistician*, 27(3/4):217–235, September - December 1978.

[74] Andreas F. Molisch. Ultrawideband propagation channels-theory, measurement, and modeling. *IEEE Transactions on Vehicular Technology*, 54:1528–1545, September 2005.

[75] Saeed S. Ghassemzadeh, Rittwik Jana, Christopher W. Rice, William Turin, and Vahid Tarokh. A statistical path loss model for in-home UWB channels. *IEEE Conference on Ultra Wideband Systems and Technologies, UWBST*, pages 59–64, May 2002.

[76] John D. Parsons. *The Mobile Radio Propagation Channel*. John Wiley & Sons LTD, 2nd edition, 2000.

[77] Thomas Zasowski, Gabriel Meyer, Frank Althaus, and Armin Wittneben. Propagation effects in UWB body area networks. *IEEE International Conference on Ultra-Wideband, ICU 2005*, pages 16–21, September 2005.

[78] Thomas Zasowski, Gabriel Meyer, Frank Althaus, and Armin Wittneben. UWB signal propagation at the human head. *IEEE Transactions on Microwave Theory and Techniques*, 54(4):1846–1857, April 2006.

[79] SEMCAD, Simulation Platform for Electromagnetic Compatibility, Antenna Design, and Dosimetry. [Online].http://www.semcad.com.

[80] Hans H. Meinke and Friedrich-Wilhelm Gundlach. *Taschenbuch der Hochfrequenztechnik, Band 1*. Springer-Verlag, 5th edition, 1992.

[81] FCC. Tissue dielectric properties calculator. Tissue Dielectric Properties Calculator, [Online].http://www.fcc.gov/fcc-bin/dielec.sh, based on results from "Compilation of the Dielectric Properties of Body Tissues at RF and Microwave Frequencies" by Camelia Gabriel, Brooks Air Force Technical Report AL/OE-TR-1996-0037.

[82] Thomas Zasowski, Frank Althaus, and Armin Wittneben. Temporal cognitive UWB medium access in the presence of multiple strong signal interferers. *14th IST Mobile & Wireless Communications Summit*, June 2005.

[83] James S. McLean, Heinrich Foltz, and Robert Sutton. Pattern descriptors for UWB antennas. *IEEE Transactions on Antennas and Propagation*, 53(1):553–559, January 2005.

[84] Samuel Dubouloz, Benoît Denis, Sebastien de Rivaz, and Laurent Ouvry. Performance analysis of LDR UWB non-coherent receivers in multipath environments. *IEEE International Conference on Ultra-Wideband, ICU 2005*, pages 491–496, September 2005.

[85] Federal Office of Communication, OFCOM, Location of Radio Transmitters. [Online]. Available: http://www.funksender.ch/webgis/bakom.php?lang=en, June 25, 2007.

[86] European Telecommunications Standards Institute (ETSI). Digital cellular telecommunications system (phase 2); radio transmission and reception (GSM 05.05 version 4.22.2), ETS 300 577. Global System for Mobile Communications (GSM), December 1998.

[87] European Telecommunications Standards Institute (ETSI). Digital cellular telecommunications system (phase 2); multiplexing and multiple access on the radio path (GSM 05.02 version 4.10.1), ETS 300 574. Global System for Mobile Communications (GSM), August 1999.

[88] European Telecommunications Standards Institute (ETSI). Digital cellular telecommunications system (Phase 2+); Multiplexing and multiple access on the radio path. 3GPP TS 45.002 version 6.12.0 Release 6, November 2005.

[89] European Telecommunications Standards Institute (ETSI). Digital enhanced cordless telecommunications (DECT); common interface (CI); part 2: Physical layer (PHL). ETSI EN 300 175-2, V.1.7.1, July 2003.

[90] European Telecommunications Standards Institute (ETSI). Digital enhanced cordless telecommunications (DECT); common interface (CI); part 3: Medium access control (MAC) layer. ETSI EN 300 175-3, V.1.7.1, July 2003.

[91] Bluetooth Special Interest Group (SIG). Specification of the bluetooth system, version 1.2, November 2003.

[92] Philip E. Gawthrop, Frank H. Sanders, Karl B. Nebbia, and John J. Sell. Radio spectrum measurements of individual microwave ovens, volume 1. National Telecommunications and Information Administration (NTIA), Report 94-303-1, March 1994.

[93] IEEE 802.11 Working Group. Supplement to IEEE standard for information technology, Part 11: Wireless LAN medium access control (MAC) and physical layer (PHY) specifications: Higher-speed physical layer extension in the 2.4 GHz band. IEEE Std 802.11b-1999 (Supplement to ANSI/IEEE Std 802.11, 1999 Edition), September 1999.

[94] IEEE 802.11 Working Group. Part 11: Wireless lan medium access control (mac) and physical layer (phy) specifications. IEEE Std 802.11-1997, June 1997.

[95] European Telecommunications Standards Institute (ETSI). Universal mobile telecommunications system (UMTS); physical layer - general description. 3GPP TS 25.201 version 6.1.0 Release 6, December 2004.

[96] European Telecommunications Standards Institute (ETSI). Universal mobile telecommunications system (UMTS); UE radio transmission and reception (FDD). 3GPP TS 25.101 version 6.7.0 Release 6, March 2005.

[97] Bundesamt für Kommunikation (BAKOM). Faktenblatt UMTS. Version 2.2, November 2004.

[98] European Telecommunications Standards Institute (ETSI). Universal mobile telecommunications system (UMTS); physical channels and mapping of transport channels onto physical channels (FDD). 3GPP TS 25.211 version 6.4.0 Release 6, December 2004.

[99] European Telecommunications Standards Institute (ETSI). Universal mobile telecommunications system (UMTS); Multiplexing and channel coding (FDD). 3GPP TS 25.212 version 6.4.0 Release 6, March 2005.

[100] UMTS TDD Alliance, Deployments. [Online]. Available: http://www.umtstdd.org/deployments.html, January 22, 2007.

[101] European Telecommunications Standards Institute (ETSI). Universal mobile telecommunications system (UMTS); UE radio transmission and reception (TDD). 3GPP TS 25.102 version 6.0.0 Release 6, December 2003.

[102] European Telecommunications Standards Institute (ETSI). Universal mobile telecommunications system (UMTS); physical channels and mapping of transport channels onto physical channels (TDD). 3GPP TS 25.221 version 6.3.0 Release 6, March 2005.

[103] Florian Trösch, Frank Althaus, and Armin Wittneben. Pulse position pre-coding exploiting UWB power constraints. *IEEE 6th Workshop on Signal Processing Advances in Wireless Communications, SPAWC*, pages 395–399, June 2005.

[104] Ralph Hoctor. Multiple access capacity in multipath channels of delay-hopped transmitted-reference UWB. *IEEE Conference on Ultra Wide Band Systems and Technologies, UWBST*, pages 315–319, November 2003.

[105] Stefan Franz and Urbashi Mitra. On optimal data detection for UWB transmitted reference systems. *IEEE Global Telecommunications Conference, Globecom*, 2:744–748, December 2003.

[106] Liuqing Yang and Georgios B. Giannakis. Optimal pilot waveform assisted modulation for ultrawideband communications. *IEEE Transactions on Wireless Communications*, 3(4):1236–1249, July 2004.

[107] Nikola K. Stanchev and Dobri M. Dobrev. Transmitted-reference quaternary phase modulation scheme for impulse radio. *International Conference on Computer as a Tool, Eurocon*, 2:1807–1809, November 2005.

[108] Dirk Dahlhaus. Post detection integration for ultra-wideband systems with binary orthogonal signaling. *IEEE International Conference on Ultra-Wideband, ICU 2005*, pages 136–141, September 2005.

[109] Thomas Zasowski and Armin Wittneben. UWB transmitted reference receivers in the presence of co-channel interference. *The 17th Annual IEEE International Symposium on Personal, Indoor and Mobile Radio Communications, PIMRC*, September 2006.

[110] Thomas Zasowski, Florian Trösch, and Armin Wittneben. Partial channel state information and intersymbol interference in low complexity UWB PPM detection. *IEEE International Conference on Ultra-Wideband, ICUWB 2006*, pages 369–374, September 2006.

[111] Athanasios Papoulis and S. Unnikrishna Pillai. *Probability, Random Variables and Stochastic Processes*. McGraw-Hill Higher Education, 4th edition, 2002.

[112] Martin Weisenhorn and Walter Hirt. ML receiver for pulsed UWB signals and parital channel state information. *IEEE International Conference on Ultra-Wideband, ICU 2005*, pages 379–384, September 2005.

[113] Florian Trösch, Frank Althaus, and Armin Wittneben. Modified pulse repetition coding boosting energy detector performance in low data rate systems. *IEEE International Conference on Ultra-Wideband, ICU 2005*, pages 508–513, September 2005.

[114] Romeo Giuliano and Franco Mazzenga. On the coexistence of power-controlled ultrawide-band systems with UMTS, GPS, DCS1800, and fixed wireless systems. *IEEE Transactions on Vehicular Technology*, 54:62–81, January 2005.

[115] Didier Landi and Christian Fischer. The effects of UWB interference on GSM systems. *2004 International Zurich Seminar on Communications, IZS*, pages 86–89, February 2004.

[116] Jeffrey R. Foerster. Interference modeling of pulse-based UWB waveforms on narrowband systems. *IEEE 55th Vehicular Technology Conference, VTC Spring*, 4:1931–1935, May 2002.

[117] Ananthram Swami, Brian Sadler, and Joi Turner. On the coexistence of ultra-wideband and narrowband radio systems. *IEEE Military Communications Conference, MILCOM*, 1:16–19, October 2001.

[118] Matti Hämäläinen, Raffaello Tesi, and Jari Iinatti. UWB co-existence with IEEE802.11a and UMTS in modified Saleh-Valenzuela channel. *International Workshop on Ultra Wideband Systems joint with Conference on Ultrawideband Systems and Technologies, Joint UWBST & IWUWBS*, pages 45–49, May 2004.

[119] Matti Hämäläinen, Raffaello Tesi, and Jari Iinatti. On the UWB system performance studies in AWGN channel with interference in UMTS band. *IEEE Conference on Ultra Wideband Systems and Technologies, UWBST*, pages 321–325, May 2002.

[120] Babak Firoozbakhsh, Thomas G. Pratt, and Nikil Jayant. Analysis of IEEE 802.11a interference on UWB systems. *IEEE Conference on Ultra Wideband Systems and Technologies, UWBST*, pages 473–477, November 2003.

[121] Kohei Ohno, Takanori Ikebe, and Testushi Ikegami. A proposal for an interference mitigation technique facilitating the coexistence of bi-phase UWB and other wideband systems. *International Workshop on Ultra Wideband Systems joint with Conference on Ul-*

trawideband Systems and Technologies, Joint UWBST & IWUWBS, pages 50–54, May 2004.

[122] Tetsushi Ikegami and Kohei Ohno. Interference mitigation study for uwb impulse radio. *The 14th IEEE 2003 International Symposium on Personal, Indoor and Mobile Radio Communciation Proceedings, PIMRC*, 1:583–587, 2003.

[123] Jim Lansford. UWB coexistence and cognitive radio. *International Workshop on Ultra Wideband Systems joint with Conference on Ultrawideband Systems and Technologies, Joint UWBST & IWUWBS*, pages 18–21, May 2004.

[124] Jim Lansford. DAA for multi-band OFDM UWB. *IEEE 802.15-05-0575-00-003a*, September 2005.

[125] Joe Decuir. Simple DAA support. *IEEE 802.15-06-0047-00-004a*, January 2006.

[126] Thomas Zasowski and Armin Wittneben. Performance of UWB systems using a temporal detect-and-avoid mechanism. *IEEE International Conference on Ultra-Wideband, ICUWB 2006*, pages 495–500, September 2006.

[127] Leonard Kleinrock and Fouad A. Tobagi. Packet switching in radio channels: Part I–Carrier sense multiple-access modes and their throughput-delay characteristics. *IEEE Transactions on Communications*, 23(12):1400–1416, December 1975.

[128] Andrew S. Tanenbaum. *Computer Networks*. Prentice Hall, 4th edition, 2002.

[129] Norman Abramson. Development of the ALOHANET. *IEEE Transactions on Information Theory*, 31(2):119–123, March 1985.

[130] Cihan Tepedelenlioğlu, Ali Abdı, and Georgios B. Giannakis. The riccan K factor: Estimation and performance analysis. *IEEE Transactions on Wireless Communications*, 2(4):799–810, July 2003.

Curriculum Vitae

Name: **Thomas Zasowski**
Date of Birth: December 14, 1975
Birthplace: Villach, Austria

Education

03/2002 –   **ETH Zurich, Zurich, Switzerland**
06/2007   PhD studies in Electrical Engineering at ETH Zurich, Communication
             Technology Laboratory.

10/1996 –   **Saarland University, Saarbrücken, Germany**
02/2002   Studies in Electrical Engineering with major in communication technol-
             ogy and microelectronics, degree: Diplom-Ingenieur

08/1986 –   **Willi-Graf-Gymnasium, Saarbrücken, Germany**
06/1995   Abitur

Experience

03/2002 –   **ETH Zurich, Zurich, Switzerland**
06/2007   Scientific assistant at the Communication Technology Laboratory with
             Prof. Dr. A. Wittneben

             → Performed research in the area of physical and medium access layer
               for ultra wideband body area networks

             → Managed and worked on two research projects with Phonak AG,
               Stäfa, Switzerland

             → Supervised diploma- and semester theses

             → Gained teaching experience

             → Reviewed international conference contributions and journals

             → Given presentations at international conferences

11/2000 –   **Ascom ART, Mägenwil, Switzerland**
02/2001   → Implemented image processing algorithms

03/1998 –   **Saarland University, Saarbrücken, Germany**
06/2000   → Supervised the VHDL laboratory at the chair of microelectronics

             → Prepared lectures at the chair of microelectronics

Publications

Florian Troesch, Christoph Steiner, Thomas Zasowski, Thomas Burger, and Armin Wittneben. Hardware Aware Optimization of an Ultra Low Power UWB Communication System. *2007 IEEE International Conference on Ultra-Wideband, ICUWB 2007.*

Thomas Zasowski, Florian Troesch, and Armin Wittneben. Partial Channel State Information and Intersymbol Interference in Low Complexity UWB PPM Detection. *2006 IEEE International Conference on Ultra-Wideband, ICUWB*, pages 369–374, September 2006, (invited paper).

Thomas Zasowski and Armin Wittneben. Performance of UWB Systems using a Temporal Detect-and-Avoid Mechanism. *2006 IEEE International Conference on Ultra-Wideband, ICUWB*, pages 495–500, September 2006.

Thomas Zasowski and Armin Wittneben. UWB transmitted reference receivers in the presence of co-channel interference. *The 17th Annual IEEE International Symposium on Personal, Indoor and Mobile Radio Communications, PIMRC*, September 2006.

Thomas Zasowski, Gabriel Meyer, Frank Althaus, and Armin Wittneben. UWB signal propagation at the human head. *IEEE Transactions on Microwave Theory and Techniques*, 54:1846–1857, April 2006.

Thomas Zasowski, Gabriel Meyer, Frank Althaus, and Armin Wittneben. Propagation effects in UWB body area networks. *IEEE International Conference on Ultra-Wideband, ICU 2005*, pages 16–21, September 2005.

Thomas Zasowski, Frank Althaus, and Armin Wittneben. Temporal cognitive UWB medium access in the presence of multiple strong signal interferers. *14th IST Mobile & Wireless Communications Summit*, June 2005.

Thomas Zasowski, Frank Althaus, and Armin Wittneben. An energy efficient transmitted-reference scheme for ultra wideband communications. *International Workshop on Ultra Wideband Systems joint with Conference on Ultrawideband Systems and Technologies, Joint UWBST & IWUWBS*, pages 146–150, May 2004.

Thomas Zasowski, Frank Althaus, Mathias Stäger, Armin Wittneben, and Gerhard Tröster. UWB for noninvasive wireless body area networks: Channel measurements and results. *IEEE Conference on Ultra Wideband Systems and Technologies, UWBST*, pages 285–289, November 2003.

Frank Althaus, Thomas Zasowski, and Armin Wittneben. Path-diversity for phase detection in low-cost sensor networks. *IEEE Signal Processing Advances in Wireless Communications, SPAWC 2003*, pages 175–179, June 2003.